D0930185

ATHLON

Loris Shano Russell, B.SC., M.A., PH.D., LL.D., F.R.S.C.

ROYAL ONTARIO MUSEUM
LIFE SCIENCES
MISCELLANEOUS PUBLICATIONS

ATHLON

Essays on Palaeontology in Honour of Loris Shano Russell

Edited by C. S. Churcher

ROYAL ONTARIO MUSEUM
PUBLICATIONS IN LIFE SCIENCES

The Royal Ontario Museum publishes three series in the Life Sciences:

LIFE SCIENCES CONTRIBUTIONS, a numbered series of original scientific publications, including monographic works.

LIFE SCIENCES OCCASIONAL PAPERS, a numbered series of original scientific publications, primarily short and usually of taxonomic significance.

LIFE SCIENCES MISCELLANEOUS PUBLICATIONS, an unnumbered series of publications of varied subject matter and format.

All manuscripts considered for publication are subject to the scrutiny and editorial policies of the Life Sciences Editorial Board, and to review by persons other than Museum staff who are authorities in the particular field involved.

©The Royal Ontario Museum, 1976
100 Queen's Park, Toronto
ISBN: 0-88854-157-0
PRINTED AND BOUND IN CANADA AT THE UNIVERSITY OF TORONTO PRESS

Contents

Preface

The opportunity to be involved in the production and editing of Loris Shano Russell's essays was one that I welcomed. It has provided me with the chance to show in a manner other than the contribution of a scientific paper my appreciation of Loris' help and influence on my career as a palaeontologist and scientist after my arrival in Canada. Loris' name was known to me before I registered at the University of Toronto in 1954, and I had come to this university in the hope of becoming his student. However, news travelled slowly and I arrived in Toronto after he had taken up the position of Chief of the Zoology Section of the National Museum of Canada in Ottawa. I was fortunate to meet Loris twice during his tenure at the National Museum, but did not get to know him. His return to the Royal Ontario Museum in 1963 was therefore the beginning of my real acquaintance and developing friendship with him, and it has been largely through his influence and the opportunities that he has opened for me that I have been able to investigate so much of the Pleistocene deposits of the Canadian prairies. Loris has taken me to many sites that he has known intimately for many years, has discoursed knowledgeably on palaeontology, stratigraphy, and local customs and history, and has instructed me in many aspects of the practice of the inter-disciplinary subject called "Vertebrate Palaeontology". I was frustrated in my earlier endeavour to become his graduate student, but now count myself one of his unregistered and unofficial students and continue to learn from my association with him.

The papers collected together in this volume do not form a scientific whole but do reflect the wide range of palaeontological interests and friendships that Loris has formed and maintained. Many specific fields are involved, including stratigraphy, invertebrate palaeontology, and history, and Loris has interests in all of those. The contributors and editors of this volume are therefore united in their wish to express their high regard for Loris, regardless of their own fields of interest or participation in this volume, and are one in heart with others who regretfully had to decline the invitation to participate in the volume because of untoward circumstances. To me personally, the editing of this volume has been a task willingly undertaken in honour of a man whom I greatly admire.

C.S. Churcher
Professor, Department of Zoology, University of Toronto, and Research Associate, Department of Vertebrate Palaeontology, Royal Ontario Museum.

Loris Shano Russell:
An Appreciation

W.E. Swinton
Massey College, University of Toronto

The German word *Festschrift* has fortunately come into the English language, or at least into its most modern dictionaries. It means literally "a writing in honour", and, as the following pages show, the writings are contributed to honour Dr. Loris Russell. They say hail but not farewell; they do not form a eulogy, at least not intentionally, but are tributes from colleagues and friends, from readers and writers, and from fellow workers in the many fields of study that Dr. Russell has helped to inaugurate or to enlighten.

Some of the papers are original contributions to sciences or arts that have particularly interested the man honoured by this book. The Appreciation is, however, the place for the facts of Dr. Russell's life, his successes, and his honours, and in it one can say things that are denied to the writers of more original texts.

Loris Russell was born in Brooklyn, New York, in April 1904. Shortly afterwards, in 1908, his parents moved to Calgary, and it was here that the future palaeontologist and historian went to public and high schools. In 1927 he graduated at the University of Alberta, as Bachelor of Science in Geology.

His postgraduate studies were undertaken in the United States at Princeton, where he received his M.A. and in 1930 his PH.D., the latter specifically in Palaeontology. Much of his subsequent career must have been determined by the inspiration of his teacher, William Berryman Scott, the great Princeton geologist and palaeontologist. Here, without doubt, Loris Russell's feet were set on a pathway, from which he has occasionally strayed, but always keeping to the sound principles in which he had been taught.

His graduation from Princeton saw him return to Canada, to another kind of school, a rigorous and demanding one, for he served seven years as Assistant Geologist with the Geological Survey of Canada. The experience was exactly what was needed to weld the ideas of a young and energetic geologist and palaeontologist, but Loris Russell was not, by choice or fate, to be only a field geologist. Academic life called, and in this milieu he was to find years of productive opportunity and fulfilment. He joined the Royal Ontario Museum of Palaeontology, then one of the five constituents of that famous university museum, and was Assistant Director from 1937 until 1940 and Director from 1946 to 1950.

These were busy years, in which Dr. Russell maintained and supplemented a famous collection, combining this work with the teaching of Palaeontology in the University of Toronto Department of Geology, first as Assistant Professor from 1937 to 1947, and then as Associate Professor from 1947 to 1950. Academic work was interrupted from 1942 to 1945 by service with the Canadian Army, where Loris, who was a radio-ham, found a valued place in the Royal Canadian Corps of Signals. He was transferred to the Reserve in 1945 with the rank of Major.

When the war was over, he was ready for museum and academic duties of a higher order, so it was not surprising that he left Toronto in 1950 for the larger field of responsibilities in the National Museum of Canada in Ottawa. He served as Chief of the Zoology Section until 1957 when he became Director of the Museum, a post which he held with distinction until 1963. Meantime the University of Alberta had recognized his work with the honorary degree of LL.D. in 1958.

Changing circumstances in Ottawa and agreeable opportunities at the Royal Ontario Museum in Toronto in 1963 were such that Dr. Russell could hardly refuse the invitation to become Head of Life Sciences or, as the post was soon renamed, Chief Biologist. Here the daily tasks were mainly of his own choosing; he could pursue palaeontology as and when he liked. Cross-appointment as Professor of Geology in the University added the stimulus of graduate students to the opportunities for field work and periods of travel.

He was now relatively free from the administrative burdens of the previous years, but he found time, as he always had, for service with other organizations. He was elected a Fellow of the Royal Society of Canada in 1936 and served as President of Section IV (Geology) from 1958 to 1959, receiving the Willet G. Miller medal of the Society in the latter year. He had long been interested in the improvement of Canada's museums and in the creation of professionally qualified staffs. As President of the Canadian Museums Association from 1961 to 1963, he had the opportunity to speak as one with authority on this field, an authority strengthened by the fact that he is one of the very few Canadians who hold the Diploma of the Museums Association (of the United Kingdom) and who achieved Distinction in the examination. I remember very clearly the success of meetings of the Canadian Museums Association in Calgary, during which Dr. Russell gave an evening discourse on the history of collecting vertebrates in Canada, and it is a pleasure to record that in 1966 much of this valuable material was published as a Royal Ontario Museum publication, *Dinosaur Hunting in Western Canada* (Contribution No. 70, Life Sciences Division, Royal Ontario Museum, U. of Toronto). More recently Dr. Russell was elected to the newly instituted Fellowship of the Canadian Museums Association and in 1972 concluded a successful year as President of the Royal Canadian Institute. Also, in August 1972 he was elected President of the International Palaeontological Union for a four-year term. Meantime he remains Curator Emeritus at the Royal Ontario Museum, a teacher and researcher, busy in winter with his select (and fortunate) band of graduate students and occupying the summers with intensive field work in Alberta and Saskatchewan.

This field work has been successful to the point where invention is required to deal with the multiplicity of small mammalian fragments. Once again the perceptive eye and the adaptive hands have enabled Dr. Russell to make the mechanical aids that greatly help his work of finding, sorting, and identifying the precious fragments of the history of Canada's past.

Fossils are many different things to many different people. First, they are objects and, like other objects, they can be collected, bought and sold, and classified. As objects they can be described, measured, photographed, and the whole series of results can be published. These results might convey some advance or increase in knowledge to a few experienced and perceptive readers.

The value of the object is greatly increased if its locality, in both the geological and geographical sense, is carefully observed and recorded. Indeed objects, whether of art or of science, are of least value for research when their provenance is neglected, as in careless or ignorant collecting it sometimes is, or if the relevant details are lost in transit or unwrapping.

Given the necessary facts, photography, and other records, fossils become more than objects; they become keys that in competent hands can unlock the past in true sequence. They can then be used for true research, and bear upon issues in both pure and applied science.

Such a fossil or series of fossils can be examined anatomically, physiologically and systematically. To all of these aspects Loris Russell has made significant contributions. This is evident from the list of his publications and need not be dealt with in depth here. But the vertebrate palaeontologist is aware of the value that constructive analysis has in the interpretation of his finds. He recreates the whole dinosaur or the tiny mammal from its parts. The study of its skull reveals the nature of the teeth, the directions of jaw movement, the kinesis or akinesis of cranial structure and, so far as it is possible, the nature of the missing brain from the empty brain-case. The skeleton is brought together piece by piece and the story of its soft parts and habits recreated. This Dr. Russell has done for both the hadrosaurs and the ceratopsians, groups for which he has extensive knowledge of their structure and horizons. Indeed they have been put in proper stratigraphical order largely through his efforts and publications.

The reconstruction of bone, muscle and skin is based on comparative anatomy. The reconstitution or restoration of the living appearance calls also for physiological understanding as well as artistry. Restoration of the moving animal poses still further problems that demand mechanical as well as physiological understanding. The final picture requires a knowledge of the environment in which the animal lived millions of years ago, whether on land, in water or in the air, what food sources were available and the climatic conditions. So the palaeontologist is anatomist, physiologist, systematist, botanist, geologist, geographer and meteorologist. If this all sounds a little too much, there is plenty of literature to dispel the doubts.

Some years ago Loris Russell decided that dinosaurs were warm-blooded and published his reasons. Today others are rediscovering the ideas, though not always for the same reasons. Today, Dr. Russell is elucidating the evolutionary pathway of small mammals. Piece by piece and by each meticulous observation the truth is becoming known. This is the work of the palaeontologist, ever conscious that we are the inheritors as well as the explorers and the explainers of evolution.

The catalogue of the fields of palaeontological knowledge or expertise given above omitted one important aspect, but one that is surely implied, that of history.

The successful palaeontologist is a historian. This aspect has never been forgotten by the recipient of these essays. Loris knows the history of the fields in which he works; he has traced the discovery and nature of oils, which are fossil fuels, and their use in letting the people of the world have light. His *Heritage of Light* tells that story: since then he has become engrossed in the simple homes that enjoyed the discovery of kerosene by Abraham Gesner, a Canadian with a London MD. This has led Dr. Russell into the byways of Canadians and antiquities generally and to Membership of the Metropolitan Toronto Conservation Authority. In all this he has applied the scientific principles used in his geologic work; here, too, they enhance his studies.

Many years ago Sir Archibald Geikie, the Scottish geologist and Director of the Geological Survey of Britain, was asked "What qualities are necessary for a geologist?" The great man replied, "A good pair of legs." Those who know the long history of geology know many examples of tenacity on legs and stout hearts on steep hills. Some of us can picture Robert Dick, the baker, geologist and botanist, who did so much for natural history, and actually on occasion walked and ran as much as 50 miles on a weekend to look at the make of the countryside and the plants of Caithness in Scotland.

Such heroic feats probably disappeared with the coming of the automobile, though agility is still called for when "bad lands" are reached or the road defeats even the modern car. Loris Russell knows the ardours and the pleasures of both kinds of travel but the automobile and the trailer have added to his mobility: he has driven long and weary roads over many years. The car has enabled him to take his wife along on all his journeys and the work of Grace Russell in the field is not to be forgotten or overlooked in any record of Loris' life and work.

In *The Doctor's Dilemma* Shaw says "all professions are conspiracies against the laity." Surely that of the palaeontologist is an exception, for most of the results of this work are displayed, and much of the research itself is done, in museums, which provide a window into far-off fields and give knowledge that is hard to come by except for the scientist. It may be noted that Dr. Russell has spent a large part of his professional life in great museums. He has thus worked for many others than himself, and most of them unknown.

Loris Shano Russell

Curriculum Vitae

BIOGRAPHICAL	Born, Brooklyn, N.Y., 1904. Married Grace Evelyn LeFeuvre, 1938.
EDUCATION	University of Alberta (B.SC., 1927) Princeton University (M.A., 1929; PH.D., 1930)
POSITIONS AND APPOINTMENTS	Research Council of Alberta, Edmonton, Alberta – Student Assistant Summers, 1925–29

Geological Survey, Department of Mines, Ottawa –
Assistant Palaeontologist — 1930–36
Assistant Geologist — 1936–37

University of Toronto –
Assistant Professor of Palaeontology — 1937–48
Assistant Director, Royal Ontario Museum of Palaeontology — 1937–45
Director, Royal Ontario Museum of Palaeontology — 1946–50
Associate Professor of Palaeontology — 1948–50

National Museum of Canada –
Chief, Zoology Section — 1950–56
Director, Natural History Branch — 1956–63
Acting Director, Human History Branch — 1958–63

University of Toronto –
Professor, Department of Geology — 1963–70
Professor Emeritus — 1970–

Royal Ontario Museum –
Head, Life Sciences Division — 1963–64
Chief Biologist — 1964–71
Curator Emeritus — 1971–

SOCIETIES

Fellow of the Royal Society of Canada
Fellow of the Geological Society of America
Fellow of the Palaeontological Society
Member, Society of Vertebrate Paleontology (President, 1958–59)
Member, American Society of Mammalogists
Member, American Ornithologists' Union
Fellow, Geological Association of Canada
Fellow, Canadian Museums Association (President, 1961–63)
Member, American Association of Museums
Member, Royal Canadian Institute (President, 1971–72)

HONOURS

LL.D., University of Alberta — 1958
Willet G. Miller Medal, Royal Society of Canada — 1959
President, Canadian Museums Association — 1961–63
Diploma with distinction, Museums Association of Great Britain — 1964

Bibliography

BOOKS

1940 Geology of the southern Alberta plains, Part 1. Stratigraphy and structure. Canada Dept. of Mines and Resources, Geol. Surv., Mem. 221. 128 pp., 21 text figs., 3 maps.

1968 A heritage of light: Lamps and lighting in the early Canadian home. Toronto, University of Toronto Press. 352 pp., 200 illus.

1973 Everyday life in colonial Canada. London, B.T. Batsford. 207 pp., 100 illus.

PAPERS

1926 A new species of the genus *Catopsalis* Cope from the Paskapoo Formation of Alberta. Amer. J. Sci., ser. 5, 12: 230–234, fig. 1.
Mollusca of the Paskapoo Formation in Alberta. Trans. Roy. Soc. Canada, ser. 3, 20(4): 207–220, pls. 1–3.

1928   A new fossil fish from the Paskapoo beds of Alberta. Amer. J. Sci., ser. 5, 15: 103–107, figs. 1–4.
       Mammal tracks from the Paskapoo beds of Alberta (with Ralph L. Lutherford). Amer. J. Sci., 15: 262–264, fig. 1.
       The genera *Kindleia* and *Stylomyleodon*. Amer. J. Sci., ser. 5, 15: 264.
       Didelphiidae from the Lance beds of Wyoming. J. Mammalogy, 9(3): 229–232, figs. 1–4.

1929   Paleocene vertebrates from Alberta. Amer. J. Sci., ser. 5, 17: 162–178, figs. 1–5.
       The validity of the genus *Stylomyleodon*. Amer. J. Sci., ser. 5, 17: 369–371.
       Upper Cretaceous and Lower Tertiary Gastropoda from Alberta. Trans. Roy. Soc. Canada, ser. 3, 23(4): 81–90, pl. 1.

1930   Upper Cretaceous dinosaur faunas of North America. Proc. Amer. Philos. Soc., 69: 133–159, table 1, fig. 1.
       A new species of *Aspideretes* from the Paskapoo Formation of Alberta. Amer. J. Sci., ser. 5, 20: 27–32, figs. 1–3.
       Early Tertiary mammal tracks from Alberta. Trans. Roy. Canad. Inst., 17(2): 217–221, pls. 7–11.
       Mollusca from the Upper Cretaceous and Lower Tertiary of Alberta. Trans. Roy. Soc. Canada, ser. 3, 25(4): 9–19, pls. 1–2.

1931   Fresh-water plesiosaurs. Canad. Field-Nat., 45: 135–137, figs. 1–5.
       Early Tertiary Mollusca from Wyoming. Bull. Amer. Paleont., 18(64): 1–38, 4 pls.

1932   New data on the Paleocene mammals of Alberta, Canada. J. Mammalogy, 13: 48–54, figs. 1–12.
       On the occurrence and relationships of the dinosaur *Troödon*. Ann. Mag. Nat. Hist., ser. 10, 9: 334–337.
       New species of Mollusca from the St. Mary River Formation of Alberta. Canad. Field-Nat., 46: 80–81, figs. 1–4.
       Stratigraphy and structure of the eastern portion of the Blood Indian Reserve, Alberta. Geol. Surv. Canada, Summ. Rept. 1931(B): 26–38, figs. 3–6.
       Mollusca from the McMurray Formation of northern Alberta. Trans. Roy. Soc. Canada, ser. 3, 26(4): 1–7, pl. 1.
       Fossil non-marine Mollusca from Saskatchewan. Trans. Roy. Canad. Inst., 18(2): 337–341, pl. 1.
       The Cretaceous-Tertiary transition of Alberta. Trans. Roy. Soc. Canada, ser. 3, 26(4): 121–156, pls. 1–2, tables 1–2.

1933   A new species of *Merychippus* from the Miocene of Saskatchewan. Canad. Field-Nat., 47: 11, figs. 1–6.
       An Upper Eocene vertebrate fauna from Saskatchewan (with R.T.D. Wickenden). Trans. Roy. Soc. Canada, ser. 3, 27(4): 53–65, fig. 1, pl. 1.

1934   Reclassification of the fossil Unionidae (fresh-water mussels) of western Canada. Canad. Field-Nat., 48: 1–4.
       Pleistocene and post-Pleistocene molluscan faunas of southern Saskatchewan. With description of a new species of *Gyraulus* by F.C. Baker. Canad. Field-Nat., 48: 34–37, figs. 1–14.
       New fossil fresh-water Mollusca from the Cretaceous and Paleocene of Montana. J. Washington Acad. Sci., 24(3): 128–131, figs. 1–5.
       Revision of the Lower Oligocene vertebrate fauna of the Cypress Hills, Saskatchewan. Trans. Roy. Canad. Inst., 20(1): 49–67, pls. 7–10.
       Fossil turtles from Saskatchewan and Alberta. Trans. Roy. Soc. Canada, ser. 3, 28(4): 101–110, pls. 1–6.

1935   A Middle Eocene mammal from British Columbia. Amer. J. Sci., ser. 5, 29: 54–55, figs. 1–4.

A plesiosaur from the Upper Cretaceous of Manitoba. J. Paleont., 9(5): 385–389, pls. 44–46.

Fauna of the Upper Milk River beds, southern Alberta. Trans. Roy. Soc. Canada, ser. 3, 29(4): 115–127, pls. 1–5.

Musculature and function in the Ceratopsia. Bull. Nat. Mus. Canada, 77: 39–48, figs. 1–9.

1936   Oil and gas possibilities along Milk River, southeastern Alberta. Canada Dept. of Mines and Resources, Geol. Surv., pap. 36–12: 1–24, 4 maps.

1937   New and interesting mammalian fossils from western Canada. Trans. Roy. Soc. Canada, ser. 3, 30(4): 75–80, pl. 1.

Revision of the geology of the southern Alberta plains. Canad. Mining Metall. Bull., 299: 185–196, 1 fig.

Preliminary Report, Del Bonita area, southern Alberta. Geol. Surv. Canada, pap. 37–10: 1–12, 1 map.

Preliminary Report, Geology of the vicinity of Taber, Alberta (with J.C. Sproule). Geol. Surv. Canada, pap. 37–14: 1–7, 1 map.

New non-marine Mollusca from the Upper Cretaceous of Alberta. Trans. Roy. Soc. Canada, ser. 3, 31(4): 61–66, pl. 1.

1938   Rattlesnakes in Alberta. Canad. Geog. J., 16(1): 33–41, 11 figs.

New species of Gastropoda from the Oligocene of Colorado. J. Paleont., 12(5): 505–507, 8 figs.

The skull of *Hemipsalodon grandis*, a giant Oligocene creodont. Trans. Roy. Soc. Canada, ser. 3, 32(4): 61–66, pls. 1–5.

1939   Land and sea movements in the Late Cretaceous of western Canada. Trans. Roy. Soc. Canada, ser. 3, 33(4): 81–99, figs. 1–8.

Notes on the occurrence of fossil fishes in the Upper Devonian of Maguasha, Quebec. Contrib. Roy. Ont. Mus. Palaeont., 2: 1–10, 1 pl.

1940   *Edmontonia rugosidens* (Gilmore), an armoured dinosaur from the Belly River Series of Alberta. Univ. of Toronto Studies, Geol. Ser., 43: 1–28, text figs. 1–2, pls. 1–8.

The sclerotic ring in the Hadrosauridae. Contrib. Roy. Ont. Mus. Palaeont., 3: 1–7, pls. 1–2.

Micrichnus tracks from the Paskapoo Formation of Alberta. Trans. Roy. Canad. Inst., 23(1): 67–74, pls. 1–2.

Titanotheres from the Lower Oligocene Cypress Hills Formation of Saskatchewan. Trans. Roy. Soc. Canada, ser. 3, 34(4): 89–100, pls. 1–5.

1941   *Prograngerella*, a new ancestral land snail from the Upper Cretaceous of Alberta. J. Paleont., 15(3): 309–311, figs. 1–4.

1943   Pleistocene horse teeth from Saskatchewan. J. Paleont., 17(1): 110–114, figs. 1–13.

Marine fauna of the Eastend Formation of Saskatchewan. J. Paleont., 17(3): 281–288, pls. 47–49, figs. 1–3.

1946   The crest of the dinosaur *Parasaurolophus*. Contrib. Roy. Ont. Mus. Palaeont., 11: 1–5, figs. 1–3.

Preliminary report on the stratigraphy of the Gaspé Limestone Series, Forillon Peninsula, Cap des Rosiers Township, County of Gaspé South. Quebec Dept. Mines, Bur. Geol. Surv., 195: 1–14, 1 map.

1947   A new locality for fossil fishes and eurypterids in the Middle Devonian of Gaspé, Quebec. Contrib. Roy. Ont. Mus. Palaeont., 12: 1–6, figs. 1–3.

1948   A Middle Paleocene mammal tooth from the foothills of Alberta. Amer. J. Sci., 246: 152–156, 1 pl.

The dentary of *Troödon*, a genus of theropod dinosaurs. J. Paleont., 22(5): 625–629, figs. 1–10.

Post-glacial occurrence of mastodon remains in southwestern Ontario. Trans. Roy. Canad. Inst., 27: 57–64, figs. 1–4.

1949   The relationships of the Alberta Cretaceous dinosaur *"Laosaurus" minimus* Gilmore. J. Paleont., 23(5): 518–520.
Preliminary report: The geology of the southern part of the Cypress Hills, southwestern Saskatchewan. Saskatchewan Geol. Surv., Petrol. Geol. Ser., 1: 1–56, pls. 1–7, fig. 1, 1 map.

1950   Correlation of the Cretaceous-Tertiary transition in Saskatchewan and Alberta. Bull. Geol. Soc. Amer., 61: 27–42, figs. 1–4.
The Tertiary gravels of Saskatchewan. Trans. Roy. Soc. Canada, ser. 3, 44(4): 51–59.

1951   Acanthodians of the Upper Devonian Escuminac Formation, Maguasha, Quebec. An. Mag. Nat. Hist., ser. 12, 4: 401–407, figs. 1–3.
Bobasatrania? canadensis (Lambe), a giant chondrostean fish from the Rocky Mountains. Ann. Rept. Nat. Mus. Canada 1949–50, Bull. 123: 218–224, pls. 1–2.
Age of the Front-Range Deformation in the North American Cordillera. Trans. Roy. Soc. Canada, ser. 3, 46(4): 47–67, figs. 1–9.
Land snails of the Cypress Hills and their significance. Canad. Field-Nat., 65(5): 174–175.

1952   Television and museums – an interim report. Bull. Canad. Mus. Assoc., 5(1): 15–16.
Out-of-doors museums. Bull. Canad. Mus. Assoc., 5(2): 9–12, fig. 1.
Cretaceous mammals of Alberta. Ann. Rept. Nat. Mus. Canada 1950–51, Bull. 126: 110–119, pls. 14–15.
Molluscan fauna of the Kishenehn Formation, southeastern British Columbia. Ann. Rept. Nat. Mus. Canada 1950–51, Bull. 126: 120–141, figs. 4–10, pls. 16–19.
Television and museums – second report. Bull. Canad. Mus. Assoc., 5(4): 14–16.

1953   Gettysburg, a museum of fields and hillsides. Bull. Canad. Mus. Assoc., 6(1): 10–12, fig. 1.
Upper Cretaceous stratigraphy of southwestern Saskatchewan. Billings Geol. Soc., Guidebook Fourth Annual Field Conference, pp. 87–97, figs. 1–7.
Tertiary stratigraphy of southwestern Saskatchewan. Billings Geol. Soc., Guidebook Fourth Annual Field Conference, pp. 106–113, figs. 1–5.
Museums news bulletins. Bull. Canad. Mus. Assoc., 6(2): 8–11.

1954   Evidence of tooth structure on the relationships of the early groups of Carnivora. Evolution, 8(2): 166–171, figs. 1–7.
A new species of eurypterid from the Devonian of Gaspé. Ann. Rept. Nat. Mus. Canada 1952–53, Bull. 132: 83–91, fig. 1, pls. 1–2.
Mammalian fauna of the Kishenehn Formation, southeastern British Columbia. Ann. Rept. Nat. Mus. Canada 1952–53, Bull. 132: 92–111, figs. 1–8, pls. 1–3.
The Eocene-Oligocene transition as a time of major orogeny in western North America. Trans. Roy. Soc. Canada, ser. 3, 48(4): 65–69.
A new species of *Cephalaspis* from the Devonian Gaspé sandstone at D'Aiguillon. Le Naturaliste Canadien, 81(12): 245–254, figs. 1–11.

1955   Television and museums – third report. Bull. Canad. Mus. Assoc., 8(1): 10–16.
Recollections of some Alberta museums. Bull. Canad. Mus. Assoc., 8(3): 9–12.
Additions to the molluscan fauna of the Kishenehn Formation, southeastern British Columbia and adjacent Montana. Ann. Rept. Nat. Mus. Canada 1953–54, Bull. 136: 102–119, figs. 1–7, pls. 1–3.

1956   Plastics in the museum. Bull. Canad. Mus. Assoc., 9(1): 10–15.
Additional occurrences of fossil horse remains in western Canada. Ann. Rept. Nat. Mus. Canada 1954–55, Bull. 142: 153–154.
Non-marine Mollusca from the North Park Formation of Saratoga Valley, Wyoming. J. Paleont., 30(5): 1260–1263, figs. 1–4.

The National Museum of History, Mexico. Bull. Canad. Mus. Assoc., 9(4): 15–16.

The Cretaceous reptile *Champsosaurus natator* Parks. Nat. Mus. Canada, Bull. 145: 1–51, figs. 1–6, pls. 1–12.

1957 International Geophysical Year exhibit at the National Museum. Bull. Canad. Mus. Assoc., 10(1): 8–12.

One man's impressions of the Lincoln Nebraska meeting of the American Association of Museums. Bull. Canad. Mus. Assoc., 10(2): 10–14.

The eleventh annual meeting of the Northeast Museums Conference, Montreal and Quebec Cities, September 3rd, 4th and 5th, 1957. Bull. Canad. Mus. Assoc., 10(3): 1–5.

Historical conservation along the St. Lawrence Seaway. Bull. Canad. Mus. Assoc., 10(3): 7–9.

Tertiary plains of Alberta and Saskatchewan. Proc. Geol. Assoc. Canada, 9: 17–19.

1958 Mollusca from the Tertiary of Princeton, British Columbia. Ann. Rept. Nat. Mus. Canada 1955–56, Bull. 147: 84–95, pls. 1–2.

Paleocene mammal teeth from Alberta. Ann. Rept. Nat. Mus. Canada 1955–56, Bull. 147: 96–103, pl. 1.

A horse astragalus from the Hand Hills Conglomerate of Alberta. Nat. Mus. Canada, Natur. Hist. Papers, 1: 1–3, figs. 1–2.

Report on the annual meeting of the Canadian Museums Assoc., Windsor, Ontario, May 7th to 9th, 1958. Bull. Canad. Mus. Assoc., 11(2): 3–15.

A palaeontological view of convergence, parallelism and orthogenesis. xvth International Congress of Zoology, 1(24).

1959 The dentition of rabbits and the origin of the Lagomorpha. Bull. Nat. Mus. Canada, Contrib. Zool., 1958, 166: 41–45, figs. 1–3.

Continental zoology of the North American Pleistocene. McGill University, Problems of the Pleistocene and Arctic, 1(1): 39–45.

1960 Fossil mammals and intercontinental connections. Roy. Soc. Canada, Studia Varia, Evolution: Its science and doctrine, ser. 4(3): 63–78, fig. 1.

The geological record of evolution. Roy. Soc. Canada, Studia Varia, Evolution: Its science and doctrine, ser. 4(4): 2–11.

1961 The National Museum of Canada 1910 to 1960. Dept. Northern Affairs and Nat. Res., Ottawa, pp. 1–37, figs. 1–23.

Swinton to ROM. Bull. Canad. Mus. Assoc., 13(4).

1962 Mammal teeth from the St. Mary River Formation (Upper Cretaceous) at Scabby Butte, Alberta. Nat. Mus. Canada, Natur. Hist. Pap., 14: 1–4, figs. 1–6.

Mammalian migrations in the Pleistocene. McGill University, Problems of the Pleistocene and Arctic, 2(2): 48–55.

Rudolph Martin Anderson. Canad. Geog. J., 45(6): 198–199.

1963 Canadian museums – this Centenary and the next. Bull. Canad. Mus. Assoc., 14(4).

Problems and potentialities of the history museum. Curator, 6(4): 341–349, figs. 1–2.

The coming of kerosene. Rushlight, 30(2): 3–6.

1964 Kishenehn Formation. Bull. Canad. Petrol. Geol., 12: 536–543, figs. 1–6.

Cretaceous non-marine faunas of northwestern North America. Roy. Ont. Mus. Life Sci. Contrib., 61: 1–24.

1965 The problem of the Willow Creek Formation. Canad. J. Earth Sci., 2(1): 11–14.

Pushing back the dawn. Meeting Place 1(3), *in* Varsity Graduate 11(5): 82–87.

Mammalian fossils from the Upper Edmonton Formation. *In* Vertebrate Paleontology in Alberta. Report of a conference held at the University of Alberta, Aug. 29 to Sept. 3, 1963. University of Alberta, pp. 32–40, figs. 5–9.

The continental Tertiary of western Canada. *In* Vertebrate Paleontology in Alberta.

Report of a conference held at the University of Alberta, Aug. 29 to Sept. 3, 1963. University of Alberta, pp. 41–52, figs. 10–13.

Body temperature of dinosaurs and its relationship to their extinction. J. Paleont., 39(3): 497–501.

Alice Evelyn Wilson, 1881–1964. Canad. Field-Nat., 79(3): 159–161.

Tertiary mammals of Saskatchewan, Part 1: The Eocene fauna. Roy. Ont. Mus., Life Sci. Contrib., 67: 1–33, pls. 1–7.

Macropalaeontology of the surface formations, Cypress Hills area, Alberta and Saskatchewan. Alberta Soc. Petrol. Geol. 15th Annual Field Conference Guidebook, Part 1, Cypress Hills Plateau, pp. 131–136.

The mastodon. Roy. Ont. Mus. Ser. What? Why? When? How? Where? Who? 6: 1–16, figs. 1–8.

1966 Frank Harris McLearn, 1885–1964. Proc. Roy. Soc. Canada, ser. 4, 3: 135–139, pl. 1.

A Paleocene conglomerate in westcentral Alberta. Canad. J. Earth Sci., 3(1): 127–128.

Lighting the pioneer Ontario home. Roy. Ont. Mus. Ser. What? Why? When How? Where? Who? 12: 1–16, figs. 1–22.

The changing environment of the dinosaurs in North America. Advancement of Science, 23(110): 197–204, figs. 1–3.

Dinosaur hunting in western Canada. Roy. Ont. Mus. Life Sci. Contrib. 70: 1–37, figs. 1–16.

Exploring the "New Red Sandstone": study indicates the Atlantic provinces may be a source for fossils. Meeting Place 1(8), *in* Varsity Graduate 13(1): 105–107.

1967 *Comment on* The inability of dinosaurs to hibernate as a possible key factor in their extinction, by John M. Cys. J. Paleont., 41(1): 267.

Review of Fossil mammals of the Lance Formation of Wyoming, Part II: Marsupialia. By William A. Clemens, Jr., Univ. California Publ. Geol. Sci., vol. 62, 1966, *in* J. Paleont., 41(3): 813–814.

Review of Marsh's dinosaurs. The collections from Como Bluff. By John H. Ostrom and John S. McIntosh, New Haven and London: Yale University Press, 1966, *in* J. Paleont., 41(4): 1029.

Confederation lamps. Canad. Antiq. Coll., 2(8): 9–11, figs. 1–6.

Unionidae from the Cretaceous and Tertiary of Alberta and Montana. J. Paleont., 41(5): 1116–1120, figs. 1–5.

A pedunculate cirripede from Upper Cretaceous rocks of Saskatchewan. J. Paleont., 41(6): 1544–1547, figs. 1–5.

Palaeontology of the Swan Hills area, northcentral Alberta. Roy. Ont. Mus. Life Sci. Contrib., 71: 1–31, pl. 1.

1968 A dinosaur bone from Willow Creek beds in Montana. Canad. J. Earth Sci., 5(2): 327–329, figs. 1–2.

A new cetacean from the Oligocene Sooke Formation of Vancouver Island, British Columbia. Canad. J. Earth Sci., 5(4): 929–933, figs. 1–11.

1969 Banquet lamp from the 1870's – a correction. Ontario Showcase, 4(34): 6–7, fig. 1.

Adventures in old-time lighting. Rotunda, Bull. Roy. Ont. Mus., 2(1): 16–25, figs. 1–11.

1970 Can we neglect research? Museum News, J. Amer. Assoc. Mus., 48(6): 13–14, 48.

The great Saskatchewan mouse mine. Rotunda, Bull. Roy. Ont. Mus., 3(1): 16–24, figs. 1–9.

Correlation of the Upper Cretaceous Montana Group between southern Alberta and Montana. Canad. J. Earth Sci., 7(4): 1099–1108, figs. 1–3.

1971 Those remarkable dinosaurs. Rotunda, Bull. Roy. Ont. Mus., 4(1): 4–17, figs. 1–13.

1972 Tertiary mammals of Saskatchewan, Part II: The Oligocene fauna, non-ungulate orders. Roy. Ont. Mus. Life Sci. Contrib., 84: 1–97, figs. 1–17.

Vertebrate palaeontology, Cretaceous to Recent, Interior Plains, Canada (with C.S. Churcher). International Geological Congress, 24th Session, Montreal. Guidebook, Field Excursion A21: 1–46, figs. 1–3.

Paléontologie des vertébrés du Crétacé au Récent des plaines intérieures au Canada (avec C.S. Churcher). Congr. géol. internat., 24ième session, Montréal. Livret-guide, Excursion A21: 1–46, figs. 1–3.

The fifty-million year pedigree of the horse. Presidential address. Proc. Roy. Canad. Inst., ser. 5, 19: 6–15.

1973   Geological evidence on the extinction of some large terrestrial vertebrates. Canad. J. Earth Sci., 10(2): 140–145.

1974   Mammals from the St. Mary River Formation (Cretaceous) of southwestern Alberta (with R.E. Sloan). Roy. Ont. Mus. Life Sci. Contrib., 95: 1–21, figs. 1–7.

Fauna and correlation of the Ravenscrag Formation (Paleocene) of southwestern Saskatchewan. Roy. Ont. Mus. Life Sci. Contrib., 102: 1–56, figs. 1–6.

1975   Revision of the fossil horses from the Cypress Hills Formation (Lower Oligocene) of Saskatchewan. Canad. J. Earth Sci., 12(4): 636–648.

Mammalian faunal succession in the Cretaceous System of western North America. In The Cretaceous System in the western interior of North America. Geol. Assoc. Canada, Spec. Pap. No. 13: 137–161, figs. 1–6.

1976   Pelecypods of the Hell Creek Formation (Uppermost Cretaceous) of Garfield County, Montana. Canad. J. Earth Sci., 13(2): 365–388.

The Image of Palaeontology

Loris S. Russell
Royal Ontario Museum

The late Mayor Camillien Houde of Montreal was once asked by radio broadcasters for permission to include a somewhat uncomplimentary reference to him in one of their programmes. His reputed reply was: "I don't care what you say about me as long as you pronounce my name right." This story, told in some variant of other public figures, ought to be taken to heart by palaeontologists. We can't agree whether we are "pally-ontologists" or "pale-ontologists". Even the way we spell our name, with or without the diphthong, is a matter of individual or national preference.

Because most of us work for government institutions or publicly supported universities, we ought to be concerned about what the public thinks of us, or whether they think of us at all. Our "public image", if any, must be a bit blurred, because there is frequent confusion, even in the press, between palaeontology and archaeology. Young people, however, seem to differentiate. Girls like to picture themselves as members of an archaeological "dig", uncovering the relics of ancient civilizations. Boys, in contrast, are more attracted by fossils, and dream of excavating giant dinosaurs in exotic places.

Fossils are most likely to be known to the public in places where they occur abundantly, like Cincinnati, Ohio, or where large collections are on display, as in New York City. England, the natal country of palaeontology, had its Liassic ammonites and ichthyosaurs and its Wealden dinosaurs well exposed to the public early in the nineteenth century. The life-size restorations of *Megalosaurus* and *Iguanodon*, made by Waterhouse Hawkins for the Crystal Palace Exposition of 1854, stimulated headlines in their day, helped by a banquet for distinguished persons staged in the lower half of the *Iguanodon* mould. In contemporary America, fossils were not as acceptable, and replicas of the Hawkins restorations presented to the City of New York ended in a garbage dump on orders from a fundamentalist mayor.

Being the history of life on the earth, palaeontology has often been caught between the firing lines of the evolutionists and the anti-evolutionists. The latter, exemplified by Georges Cuvier and Hugh Miller, made much of the gaps in the palaeontological record. Darwin put these apparent hiatuses in their proper perspective in 1859. Curiously, and in spite of a vast increase in our knowledge of fossils, groups of "creationists" in California and elsewhere are using the supposed incompleteness of the geological record again as an argument against evolution. They should go back and read Chapter Ten of *The Origin of Species*.

Sometimes our best friends can be our worst enemies. Plaster of Paris is probably the most useful of all materials in the collecting and preparing of fossils, but it has brought us some bad publicity. When the skeleton of the Bone Cabin *"Brontosaurus"* was mounted in the American Museum of Natural History in 1905 —the first such public display of a sauropod—Samuel Clemens is supposed to have said, "Professor Osborn has just reconstructed a 75-foot dinosaur. If the plaster had held out he would have made it 100." A kinder version of the same canard was overheard some years ago in the Dinosaur Gallery of the Royal Ontario Museum. A clerical gentleman said to his companion, "Of course these are plaster replicas. The originals are too valuable to be placed on public exhibition."

Some kind of extreme in this sort of thing was once reached in Calgary, Alberta, where a high-school teacher named William Aberhart established an evangelical church called the Prophetic Bible Institute. With the aid of a powerful radio station, "Bible Bill" propagated his brand of religion and economics, and eventually led his Social Credit party to power as the government of the Province of Alberta. But that is another story. Aberhart must have been bothered by those reports of scientists digging up the skeletons of ancient monsters in the Red Deer River valley, only 100 miles away. So he assured his listeners that it was all a fraud and that these people went to a secret place in the badlands, where they fabricated the bones from plaster of Paris to confound the true believers. C.M. Sternberg, on hearing of this, commented, "How clever we are!"

The most famous of all calcium-sulphate dinosaurs was the replica of the Pittsburgh *Diplodocus*, which was presented by Andrew Carnegie to the British Museum in 1905. Others subsequently went to museums in Berlin, Paris, Vienna, Bologna, St. Petersburg, La Plata, and Madrid. The presentations were accompanied by elaborate ceremonies, in which Mr. Carnegie and Dr. Holland participated prominently. One wonders if there was a let-down after all the publicity, when the interested public realized that the skeleton was not a real fossil, but a very ingenious plaster copy.

While on the subject of dinosaurs, let's not downgrade them. They are our best-known product, or as John Ostrom says, our bread and butter.

The first impact of palaeontologists on the American public may have occurred when they appeared in the unexplored west about the middle of the nineteenth century. The native people may have had various opinions about these strange intruders, but the only one we know is the verdict of the Sioux that F.V. Hayden was a lunatic, and therefore under the protection of Wakan Tanka. Years later, O.C. Marsh made some repayment for this unintentional courtesy to his predecessor by exposing the corruption of the government administration at the Pine Ridge Indian Agency.

Beginning in the 1870s, the western settlers came to know the fossil collectors well, and to call them bone-hunters. Later the name was extended to all geological explorers. In Alberta the corresponding term was "bone-picker", which suggests greater success than the American name. It was not easy in the early days to explain what our science was about. During the First World War, George Sternberg had leave of absence from the Geological Survey of Canada to help in the local harvest. To his temporary employer's natural question as to his regular work, George explained that he went to western Canada to find and bring back prehistoric animals for the government museum. "You must be a very brave man", said his host, "to capture such big, ferocious creatures." Years later, when George was at the University of Alberta, Prof. William Rowan, a distinguished zoologist and a gifted artist, based a cartoon on this incident, showing George leading a reluctant dinosaur by a lariat into a Canadian Pacific freight car.

Unfortunate publicity has sometimes been the fault of the palaeontologists themselves. Cuvier had some basis for his "law of the correlation of parts", but as years went on this became exaggerated until the public was asked to believe that "from a single small bone a scientist can reconstruct a complete skeleton". So he can, if the scientist already knows the skeleton, but it sounded so preposterous that the creditability of palaeontology became strained. Then there was the famous Cope-Marsh feud, which culminated in the washing of dirty linen in the *New York Herald* of January, 1890. The geological world was shocked, but the public must have thought the whole thing silly. The greatest publicist that palaeontology ever had was Roy Chapman Andrews, but one of his stunts backfired. To gain financial support and to publicize the Central Asiatic Expedition of the American Museum of Natural History, Andrews "auctioned" a specimen from the first clutch of dinosaur eggs brought back from the Djadochta cliffs of Mongolia. The large sum paid for this ancient egg was announced with fanfare, but the Mongolian authorities were furious, for Andrews had assured them that the Expedition was not collecting anything of monetary value. Negotiations for a renewal of the Expedition's permit were made much more difficult.

Some impressions of palaeontology have been created by novels. Jules Verne was the first to use such a theme. His explorers in *A Journey to the Centre of the Earth* encountered live ichthyosaurs in a subterranean ocean. The most famous story of this sort is Conan Doyle's *The Lost World*, the fictional narrative of a British expedition to South America, where on an isolated plateau they found survivors of the Mesozoic fauna. These and similar accounts have created the impression that prehistoric man and the dinosaurs were contemporary, and this impression has been sustained by motion pictures and comic strips. Like the big lie of propaganda, there seems to be no way that this falsehood can be overtaken by the truth.

Apart from the general public, how do students of other disciplines regard palaeontology? Our science occupies the ground between geology and biology and is open to evaluation by both the earth scientists and the life scientists. Today there seems to be a greater emphasis in geology on the physical rather than on the biological aspects of the science. In his training the average geologist gets the briefest exposure to biology and palaeontology. If the teaching is by persons not actively engaged in palaeontology it is likely to be dull, creating the impression that the subject is boring and of minor importance. W.B. Scott, in his lecture on geological correlation, used to say, "Geologists are always trying to escape from the thraldom of palaeontology." They have not succeeded, and they should remember that it was not until William Smith established the principles of biostratigraphy that geology was able to shake off the preposterous fantasies of Wernerism.

Although in theory palaeontology is a biological science, its application to biostratigraphy has tended to ally it to geology and to estrange it from biology. Most palaeontologists take their university training in departments of geology, where the emphasis is likely to be on identification and stratigraphic range. This has led biologists to regard palaeontology as a purely descriptive study, not likely to contribute much to the theory and philosophy of the life sciences. Such a view was enunciated by the geneticist T.H. Morgan, but was emphatically denied by W.D. Matthew. Genetics may be able to show how evolution could have operated, but palaeontology reveals what actually happened. Fortunately, the two approaches have been reconciled in the "synthetic" theory of evolution, and the palaeontological record can be interpreted in terms of neontological genetics. But palaeontology, especially the study of invertebrate animals, is still something of an orphan among the life sciences. Yet such groups as the Brachiopoda and the Ammonoidea are fertile fields for the study of evolution.

The bridge between palaeontology and the other biological disciplines has usually been started from the palaeontological side. Major contributions to systematics and comparative anatomy have been made by palaeontologists, and now the study of ecology is being vitalized by the palaeontological approach. Which brings me to the point of this essay, if it has one. The image of palaeontology, whether at the various public levels or in the stratosphere of the sciences, is created by the palaeontologists. Sound taxonomy and stratigraphy are basic to our science, but these must be practised with the constant awareness that we are dealing with what were once living organisms and that these still have something to tell us that is applicable to the organisms of today, including ourselves. And when we have a bright idea, let's not just leave it planted in the scientific literature, where it may or may not germinate. Dig it out for the public to see; it may serve to counterbalance some of the nonsense that is being spread around today in the guise of authentic prehistory.

A Microbioherm

Madeleine A. Fritz
Research Associate, Department of Invertebrate
Palaeontology, ROM, and Professor Emeritus of
Palaeontology, Department of Geology,
University of Toronto

Abstract

A microbioherm consisting of successive overgrowths of several species of ectoprocts and stromatoporoids was found in the holotype of the bryozoan *Stigmatella hybrida*. The sequence began on the coenosteum of a small stromatoporoid, *Dermostroma*, and ended with an extensive growth of *S. hybrida*, the presumed winner of a long competition for survival. A recent example of a biota capable of being preserved as a microbioherm is found in the sublittoral *Flustra foliacea* of the Canadian Atlantic coast. Its history would be one of symbiosis rather than competition.

Introduction

Dyer (1925) stated that the zoarium of the holotype of his species, *Stigmatella hybrida*, consisted of "a more or less cone-shaped mass". In my restudy of this type (Fritz, 1973) a thin section was prepared from a slice cut through the specimen from top to bottom. This section revealed the presence of an unsuspected rhythmical stromatoporoid – bryozoan (ectoproct) sequence in which *S. hybrida* was but one of the component elements. This assemblage was referred to briefly in my paper (Fritz, 1973). It is now described in detail and possible palaeoecological conditions under which it developed are considered. The term microbioherm, as a title, was suggested by Dr. Raul Vicencio, Research Fellow in the Department of Invertebrate Palaeontology.

Fig. 1 "Microbioherm", × 2.

Fig. 2 Polished surface of "microbioherm", slightly larger than × 2.

Description

Exterior Features

This small mound-like fossil is 65 mm wide and 35 mm thick (Fig. 1). The surface is covered with tiny, pointed monticules spaced approximately one to two millimetres apart. Two prominent tuberosities protruded originally from the surface. The one now intact has a basal diameter of 20 mm and extends 15 mm from the surface; it terminates in a rounded extremity with a shallow central concavity and suggests an aborted branch. Only the base of the other (also 20 mm in diameter) remains, Dyer having used it for the preparation of his thin sections of *Stigmatella hybrida*, unaware of the composite nature of the specimen. Other surface swellings probably represent initial stages in the growth of similar protuberances.

Internal Structure

The biotic sequence, as revealed by one thin section, (Figs. 2 and 3), began on the coenosteum of a small stromatoporoid with a convex base (Fig. 3, s 1), the upper surface of which has small mamelons and closely set papillae. It is 35 mm wide, 4 mm thick, reaching 7 mm in a mamelon. The calcareous skeleton is composed of undulating laminae connected at various intervals by pillars. The structure is that of the genus *Dermatostroma* which may have lived in the calcareous muds of the ocean floor or formed overgrowths on associated shells.

Upon the irregular surface of the stromatoporoid, with its living "hydrozoan" tissue, the bryozoan embryo of *S. hybrida* settled down and by repeated budding formed a thick overgrowth that covered practically the whole host (Fig. 3, BS 1). This overgrowth ranges in thickness from 5 to 7 mm, with an undulating upper profile; in addition, the cross-sections of two round branches 5 to 7 mm in diameter appear lower in the section (Fig. 3, BS 1. XS) indicating the presence of extensions from the bryozoan surface. The overgrowth did not kill the stromatoporoid since it appears, as will be seen, later in the assemblage. The two may have lived together in toleration or the tromatoporoid may have survived only along the periphery of the colony. However, before it appeared again an unidentified bryozoan (Fig. 3, B?) formed a laminar growth over *S. hybrida*. It is poorly preserved, and its zooecia are variously oriented, some in the normal attitude of growth, others in the reverse position suggesting that the zoarium was twisted or broken. The zooecia are relatively large with few diaphragms, thus forming an open skeletal

Fig. 3 Thin section of microbioherm showing succession of overgrowths by organisms. Abbreviations: s 1, s 2, and s 3 – first, second, and third growths of stromatoporoid, ?*Dermatostroma* sp.; BH – Bryozoan, *Heterotrypa prolifica*; BS 1, BS 2, and BS 3 – first, second, and third growths of bryozoan, *Stigmatella hybrida*; BS 1.XS – cross-sections of extensions from first growth of *Stigmatella hybrida*; B? – Unidentified bryozoan. Bar indicates 10 mm.

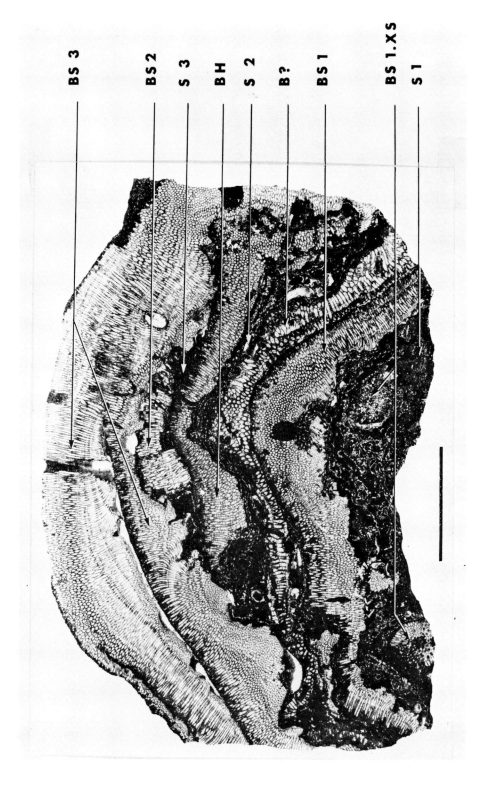

framework that responded readily to adverse conditions which temporarily
affected the colony. Overgrowing the latter, the stromatoporoid, apparently still
relatively vigorous, reappeared (Fig. 3, s 2). However, it showed signs of weaken-
ing in that the coenosteum was only 3 mm thick and 5 mm in a mamelon. Upon
the irregular surface of this second growth the bryozoan *Heterotrypa prolifica*
grew to a maximum thickness of 5 mm (Fig. 3, BH). This was, in turn, surmounted
by a third appearance of the stromatoporoid (Fig. 3, s 3), now only 1 to 2 mm
thick, making its last recognizable appearance in this unusual symbiotic series.
The sequence ended with an extensive growth of *Stigmatella hybrida* (Fig. 3, BS
3), at least 14 mm in thickness, covering the entire upper surface of the section
and in fact covering the entire specimen. It would appear that this bryozoan, which
selected the coenosteum of a species of *Dermatostroma* as a favourable place to
live, won out at the end of a long and keen competition for survival and finally
concealed all the earlier stages.

Palaeoecology

The Upper Ordovician rocks that underlie Toronto and an area some 20 miles to
the west belong to the Georgian Bay Formation (Liberty, 1969). Prior to 1969
two formations were recognized, namely the Dundas (Parks, 1924) and Meaford
(Foerste, 1924). The Meaford Formation of Richmond age was subdivided by
Dyer (1925) into three members: in ascending order — Erindale, Streetsville,
and Meadowvale. A stratigraphic section of the Erindale exposed along Cooksville
Creek about one and one-half miles north of the town of Cooksville, Ontario,
measured 24 m according to Dyer. Midway in this section a small bryozoan
biostrome ("reef" of Dyer), 0.2 to 0.5 m thick, is present. The biostrome can be
traced over at least 3 sq. miles. It is within this biostrome that the "microbioherm"
was found. In order to envisage the geological conditions under which the microbioherm grew the biostrome as a whole is considered. The rocks consist of heavy
beds of grey argillaceous limestone up to 0.3 m thick, interbedded with thin layers
of fissile shale. The limestone beds in particular carry an abundance of Bryozoa
(mostly trepostomes), numerous well-preserved brachiopods, a few molluscs, still
fewer stromatoporoids, and scattered trilobite fragments, presumably *Isotelus*.

It is generally conceded that Bryozoa (ectoprocts) develop in a calcareous mud
facies of the littoral or sublittoral zones, below drastic wave action in well-lit,
plankton-rich waters, and at depths of about 61 to 91 m. The recognizable bryozoan fauna contained in the biostrome described is represented by broken (but
not comminuted) remains of colonies which exhibited various habits of growth,
including thick overgrowths, incrustations, laminar expansions, and massive forms
of bizarre shapes. Many specimens have abraded surfaces, but most show excellent microstructure in thin sections.

To trace the long and complex history which has led to the present biostratigraphic structure is well nigh impossible. Nature is a fickle historian, recording some
events in one place only to vacate that place to continue the record elsewhere.
Then in a wanton mood she may destroy the whole record and carry the remains
into the sea. The record might be likened to a history book in which pages and

Fig. 4 *Flustra foliacea* Linnaeus, all natural size. Top – colony grown on muddy substrate; bottom left – colony grown on pebble visible at bottom right; bottom right – colony grown on debris.

whole chapters are missing, having been torn out, crumpled or destroyed by some fiend. However, it has been said that "the present is a key to the past" (see Geikie, 1905). If this be true it follows that observations on living organisms in their natural environment might be helpful in interpreting the history of their ancestors of the distant past. With this idea in mind, I am prompted to discuss the recent bryozoan *Flustra foliacea* Linnaeus (1758) (Fig. 4) which flourishes in abundance in the sublittoral zone of the Bay of Fundy at St. Martins, New Brunswick, in waters from 61 to 91 m deep. This bryozoan forms delicate calcareo-membraneous, foliaceous colonies which are fixed either to the substratum or attached to shells, pebbles, or other objects; they may also start as incrustations on shells. The colonies are torn from their off-shore moorings by the flow of the strong Fundy tides. With the ebbing of the tide, skeletons in great numbers are scattered indiscriminately on the mud flats and on the sandy beaches along with fucoids of various species. On the mud flats, in particular, the skeletal remains become broken, reworked and in the course of time buried; some are ground to dust, thus adding a calcareous element to the mud. Doubtless countless generations of *Flustra foliacea* are already buried in the calcareous mud facies and countless more will doubtless be buried in the future, providing the present geological conditions prevail.

Should the sea retreat at some future period in earth history, the calcareous muds, with their entombed remains of *Flustra* and associated fauna and flora, might be compacted in rocks and preserved, should they not be destroyed beyond recognition by the destructive forces of nature. These hypothetical rocks could constitute a bryozoan bioherm in which might be found symbiotic assemblages of Recent taxa similar to the Ordovician bioherm that contained the microbioherm which forms the subject of this paper.

Literature Cited

DYER, W.S.
   1925   The stratigraphy and paleontology of Toronto and vicinity, Pt. v, The paleontology of the Credit River section. Rep. Ont. Dept. Mines, 32(7), 1923: 47–88.

FOERSTE, A.F.
   1924   Upper Ordovician faunas of Ontario and Quebec. Geol. Surv. Canada, Mem. 138: 51.

FRITZ, M.A.
   1973   Redescription of type specimens of bryozoan *Stigmatella* from the Upper Ordovician of the Toronto region, Ontario. Life Sci. Contr., R. Ont. Mus., 87: 1–31.

GEIKIE, A.
   1905   The founders of geology, 2nd ed., p. 299.

LIBERTY, B.A.
  1969   Palaeozoic geology of the Lake Simcoe area, Ontario. Geol. Surv. Canada, Mem.
          355: 73–79.

LINNAEUS, C.
  1758   Systema naturae. Ed. 10(1). Stockholm, 824 pp.

PARKS, W.A.
  1924   Upper Ordovician at Toronto, Ontario. Bull. Geol. Soc. Am., 35: 103 (abst.).

New Species of *Ctenaspis* (Ostracodermi) from the Devonian of Arctic Canada

D.L. Dineley
Department of Geology, University of Bristol, U.K.

Abstract
Three new species of *Ctenaspis* from the Peel Sound Formation of Prince of Wales Island, N.W.T., are basically similar to known species but differ in their large size, minor details of proportions and ornamentation. The body form, tail and squamation are described for the first time and a mode of life involving partial burial in sediment is suggested.

Introduction

Relatively little is known of the scaled bodies of early Agnatha but the vertebrate faunas of the Lower Devonian Peel Sound Formation of arctic Canada provide several new examples of more or less complete heterostracan ostracoderms. A new species of *Ctenaspis* is one of these, and together with other species from Prince of Wales Island is of interest on anatomical, biostratigraphic and palaeo-ecological grounds. These species are described below and a tentative view of the mode of life of *Ctenaspis* is offered.

During an investigation of the Peel Sound Formation (Lower Devonian) on Prince of Wales Island in 1967 ostracoderm faunas were discovered at a number of localities (Broad *et al.*, 1968). The Peel Sound Formation is a conspicuous red-bed unit in the area around the Boothia Uplift (or Arch) of Prince of Wales Island, Somerset Island and the Boothia Peninsula (Thorsteinsson and Tozer, 1963; Brown *et al.*, 1969; Kerr and Christie, 1965) and comprises a wide range of lithologies (Miall, 1970a; 1970b). A facies change from fluvial conglomerates to marine carbonates occurs within the upper member of the Peel Sound Formation westwards from the vicinity of the Arch on Prince of Wales Island. The most abundant and widespread occurrences of vertebrate remains are in the sandstone-carbonate facies, especially in the pink- or grey-coloured fine sandstones and siltstones. Plant remains also are known from several of these beds. *Lingula* and fossils of marine invertebrates are present in some of the strata.

The depositional environments of the sandstone-carbonate facies are thought to include meandering and braided streams, estuaries, lagoons, deltas and cheniers adjacent to a very shallow sea. Locally there were spreads of primitive vascular plants and perhaps algal mats and thickets in the bodies of water. Water flow was variable and energy conditions generally low, with perhaps a wide coastal shelf to the west. Farther in that direction the marine carbonate facies of the Peel Sound Formation includes thin "bone beds" of comminuted vertebrate material, which no doubt were concentrated by tidal or wave action.

A fortunate consequence of the relatively quiet conditions within certain of the sandstone-carbonate facies environments is that Heterostraci have been preserved relatively intact, with shields and scales articulated, and osteichthyan remains in relatively complete assembly have also been found.

Ctenaspis has been recovered *in situ* from several localities on northern Prince of Wales Island. The lowest two (A and B) have yielded abundant material; at other localities only a few poorly preserved specimens have been collected. Localities A and B are situated on the bank of a small stream flowing north into Baring Channel (98°30'W: 73°47'N). At A is an outcrop of grey, buff-weathering, coarse- to medium-grained flaggy sandstone at the top of the eastern bank of the stream. It forms a small bluff. Locality B is on an outcrop of a 30-cm-thick bed of pink- and grey-mottled, calcareous, flaggy fine-grained sandstone-siltstone in the low cliff (about 2 m high) on the western bank some 500 m downstream from A. They are in the upper member of the Peel Sound Formation.

Associated with *Ctenaspis* at locality A are numerous headshields of a medium-sized cephalaspid, shields of an indeterminate large cyathaspid and, less commonly, a small arthrodire, *Baringaspis dineleyi* Miles (1973), and a small pteraspid. The fossils are broadly scattered on the bedding planes and, although disarticulated, are not much fragmented. It is unlikely that the articulated *Ctenaspis* remains travelled far or were much buffeted by currents, and several of the cephalaspids retain some squamation. One may conclude that the fauna was a thanatocoenose but it may represent a single local population or community.

The vertebrate-bearing stratum at locality B, however, gives a very different impression. Within the exposure along about 15 m of stream bank the cross-sections of many hundreds of ostracoderm shields were detected. In several places they are piled upon one another in imbricated fashion, the shields being those of a medium-sized pteraspid and a *Pionaspis*-like cyathaspid. On several bedding planes the fossils are very abundant, occurring as clusters or groups of several individuals with scattered individual shields between. So far no cephalaspids have been found but rare plates of *Baringaspis dineleyi* are present. The arthrodire plates are nearly complete and the *Ctenaspis*, pteraspid and cyathaspid shields are complete in most cases; no articulated specimens have been found, however, nor have associated or isolated scales. This thanatocoenose has every appearance of having passed some appreciable time in undergoing winnowing and selective transport before finally being interred. The environment in which both this assemblage, and the dolomitic silt in which it is preserved, accumulated was perhaps a tidal lagoon or estuary. No marine invertebrates here indicate the proximity

of the sea, but the fine-grained character of the rock and the sorting of the vertebrates suggest a regime where there was relatively active movement of water
but little influx of coarse sediment. The animals may have lived in these waters,
or nearby, in large numbers, but following death the bodies rapidly decomposed
and the skeletons became disarticulated without much other mechanical damage.

The other localities from which the genus has been collected are within the
lower member of the Peel Sound Formation. Loose fragments were obtained
from scree near Backe Bay and Mount Matthias, north of Browne Bay on the
east coast of Prince of Wales Island and from the southern end of Pandora Island
off the central east coast of Prince of Wales Island. The derivation of these specimens is uncertain.

Ctenaspis has hitherto been recorded in Vestspitsbergen (Kiaer, 1930),
Podolia (Zych, 1931; Stensiö, 1958), England (Wills, 1935), and Cornwallis
Island, N.W.T. (Thorsteinsson, 1958), and three species have been described.
The apparently primitive character of the armour and the distinctive shape and
ornamentation of the shields led Kiaer (1930) to erect the family Ctenaspididae.
Obruchev (1964) retained this taxon but Denison (1964) designated the
Ctenaspidinae as a sub-family within his family Cyathaspididae. In the present
account Denison's arrangement is preferred on the grounds of caution since the
author has not been able to examine all of the genera included in Obruchev's
Ctenaspididae, and because the other genera can apparently be accommodated
within the family Cyathaspididae.

There is no doubt *Ctenaspis* shows great simplification in the reduction of the
armour to two major plates, the dorsal and ventral shields. What small plates may
have existed in the vicinity of the mouth is not known. The eyes are small, but,
apart from a large pineal organ present in the second new species described below,
little can be seen of the nature of the other sensory organs. Denison (1964,
p. 326) shows a complex lateral line system in the Ctenaspidinae.

The microstructure of the shields conforms to the cyathaspid pattern but the
tubercles are capped with aspidin, rather than dentine. The thinness of the shields
is notable in the new fossils and in most individuals little of the actual material
of the shield remains, the specimens being mostly casts or moulds.

Novitskaya (1967; 1968; 1971) suggests that the amphiaspids known from
northern Siberian late Lower Devonian arose from a ctenaspid stock. This interesting idea is discussed below. Apart from possibly giving rise to the amphiaspids,
the ctenaspids seem to have been a specialised group.

In the present study extensive use has been made of latex pulls to record and
reverse the impressions of shields and scales in the rather coarse-grained matrices.
Such are the lithologies that mechanical preparation is slow, difficult and largely
unproductive, while the dolomite cement renders chemical preparation equally
unprofitable.

Systematics

Family Cyathaspidae Denison, 1964 (= suborder Cyathaspida Kiaer, 1932)
Subfamily Ctenaspidinae Denison, 1964 (Diagnosis, p. 438)

Genus *Ctenaspis* Kiaer, 1930

Diagnosis (Denison, 1964, p. 439)

Ctenaspis includes Ctenaspidinae in which the shield is short and relatively very broad (width ratio = 0.70–0.90; see Denison, 1964, p. 317, for ratios). The dorsal shield is rather flat, has the orbits far forward (orbital ratio = 0.8–0.12), and has well-developed lateral brims. The ventral shield is more strongly arched except for flatter marginal areas.

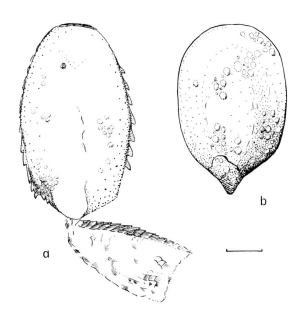

Fig. 1 *Ctenaspis obruchevi* sp. nov. Peel Sound Formation, Prince of Wales Island, N.W.T. Scale = 10 mm.
  a. Dorsal shield and scaled trunk and tail (type: NMC 21700).
  b. Ventral shield (NMC 21701).

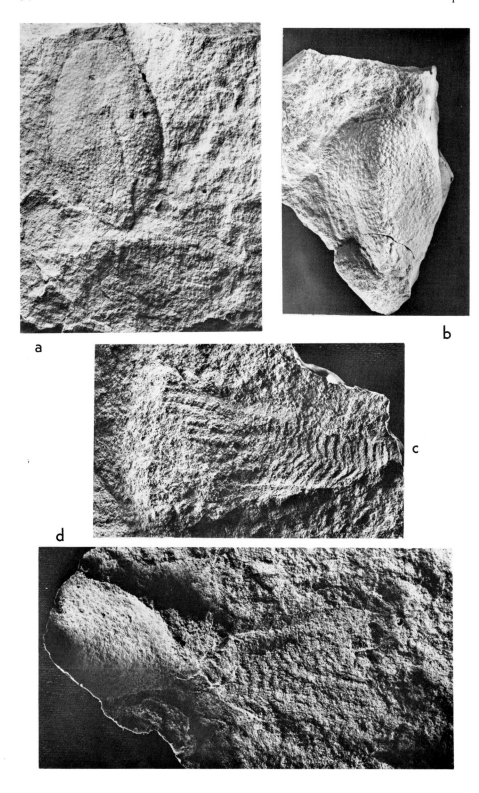

Ctenaspis obruchevi sp. nov.
Figs. 1, 2, 3

Holotype
Imperfect dorsal shield and tail (NMC 21700: National Museum of Natural Sciences, Ottawa). The species is named in honor of the late Professor D.V. Obruchev, Moscow.

Occurrence
Early Devonian (Gedinnian – ?Siegenian). Sandstone-carbonate facies of the Upper member of the Peel Sound Formation, Northern part of Prince of Wales Island, N.W.T.: Locality A.

Referred Material
NMC 21701, 21704 to 21780.

Diagnosis
A *Ctenaspis* with dorsal shield 4.00 to 5.00 cms long and with a width ratio of about 0.68, lateral brims coarsely serrate, gently arched, rostral margin conspicuous. Tubercles of ornamentation coarse, round and low. Ventral shield strongly arched with vertical lateral brims, posterior margin produced to median point, ornamentation of coarse, round tubercles.

Description
This is a relatively large species, rather more elongate than others in the same genus, with prominent coarsely serrated lateral brims to the dorsal shield. The orbits are, as is common in the genus, rather far forward, and are small. The rostral margin is distinct and smooth in outline; its dorsal surface is ornamented with fine ridges parallel to the edge. The posterior margin is broadly V-shaped and is finely serrated. There is no postero-median crest. The pineal organ is rather far back (pineal ratio = 9.25).

The pre-orbital processes and maxillary brim have not been observed. The serrate lateral margins of the dorsal shield appear to have projected laterally beyond the meeting of the dorsal and ventral shields.

The ventral shield is rather pear-shaped in outline with a relatively straight anterior margin, gently curved lateral margins and a short broadly V-shaped posterior margin. It is difficult to see the extent of the mouth as there is no obvious depression of the anterior margin to accommodate it, and no separate oral

Fig. 2. *Ctenaspis obruchevi* sp. nov. Peel Sound Formation, Prince of Wales Island, N.W.T. All figures natural size.
- a. Dorsal shield and tail, NMC 21700 (type). The tail is bent to the right at about 90° to the axis of the disc.
- b. Ventral shield, NMC 21701.
- c. Impression of scales of left flank of body and tail, NMC 21767.
- d. Cast of ventral shield, scaled body and tail, NMC 21768.

plates are known. The shield is broadest within the posterior half and is deepest in the posterior two thirds. In the posterior half of the shield the lateral quarters are flattened. The ventro-lateral lamina is almost vertical and is approximately 3 mm deep and extends to the branchial openings: it is ornamented by fine postero-ventrally directed ridges.

The surface ornamentation of the dorsal shield and the ventral shield apart from the ventro-lateral lamina is of coarse (1–2 mm diameter) rounded tubercles, smooth and in only very few instances with apices directed to the rear. Denison comments (1964, p. 440) that in *C. dentata* the tubercles are of aspidin: this is most probably true of *C. obruchevi* also.

No trace of the lateral line system has been found in this material, nor have impressions of the internal anatomy been discovered.

The Scaled Trunk and Tail (Fig. 3)

Poorly preserved trunk and tail squamation is preserved in several specimens, four of which are relatively complete.

The scaled body or trunk is short, laterally compressed and about 15 mm deep. As far as can be determined, there are four series of short thin scales: dorsal median, dorso-lateral, ventro-lateral and ventral median. Each is ornamented by fine ridges parallel to the axis of the body, the ornamentation on the median series being coarsest over the crests of the scales. There may be 12 to 15 scale-circles about the body; the degree of overlap of adjacent circles is not known. Unlike the arrangement in other cyathaspids known so far (Denison, 1964), dorso-lateral and ventro-lateral scales are about the same size, about 4 mm deep.

The tail itself comprises a deep fin which is supported from the sharply tapered body by five, or perhaps six, "rays" of conspicuous scales which may have covered endoskeleton and muscle. It is broad and symmetrical in outline. The individual scale shapes and their arrangements are unfortunately indistinct.

Discussion

The species *Ctenaspis obruchevi* is notable for the size it attains and for the relatively elongate shape of the shields. A point of interest in the specimens which retain body and tail is that these are drawn over to one side sharply so as to be almost at 90° to the axis of the shields. This may be a contractional feature acquired during or after death.

The serrations on the lateral margins of the dorsal shield are not matched by similar features on the ventral, which seems to fit snugly against the dorsal along a simple line of commisure. No branchial plates or sub-orbital plates are known and the mouth must have been exceedingly compressed, though wide.

As stated above, the shields of the Canadian species of *Ctenaspis* are very thin, and in *C. obruchevi* and the next described species there is no trace of the internal dorsal cranial anatomy except for the very large pineal swellings. It may be suggested that this relatively enlarged organ was perhaps functionally important and perhaps received stimulus directly through the bony covering above it. In an animal with small eyes and in an environment where it was perhaps frequently covered by mud or silt this organ might have added to the animal's perception of its immediate surroundings.

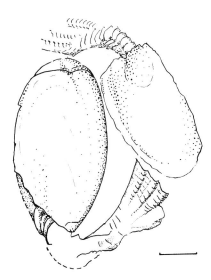

Fig. 3. *Ctenaspis obruchevi* sp. nov. Two fragmental ventral shields with associated trunk scales and tails: that on the left is a cast with the dorsal shield possibly attached on the reverse; that on the right is an external cast. The scales of the specimen on the right occur on a plane perpendicular to that of the shield but are shown above as rotated to be on the same plane. Scale = 10 mm. (NMC 21769a, 21769b.)

Ctenaspis russelli sp. nov.
Figs. 4, 5

Holotype

NMC 21702 Dorsal shield. The species is named in honour of Dr. Loris S. Russell.

Occurrence

Early Devonian (Gedinnian–?Siegenian). Sandstone-carbonate facies of the Peel Sound Formation. Northern part of Prince of Wales Island, N.W.T., Locality B.

Referred Material

NMC 21703, 21781 to 21855.

Diagnosis

A very large *Ctenaspis* with dorsal shield about 60 mm long and 50 mm wide, rostral margin with row of coarse tubercles, and lateral brims very coarsely serrate. Tubercles round and low. Ventral shield deep, lateral margins subparallel, rather square cornual angles and broadly V-shaped posterior margin: ornamentation of very coarse rounded and posteriorly directed tubercles.

Description

The dorsal shield of this large species is relatively broad, the width ratio being about 0.83. It is also relatively flat. The orbits are far forward, the orbital ratio is about 0.8; and the very large and prominent pineal is placed well back (0.23–

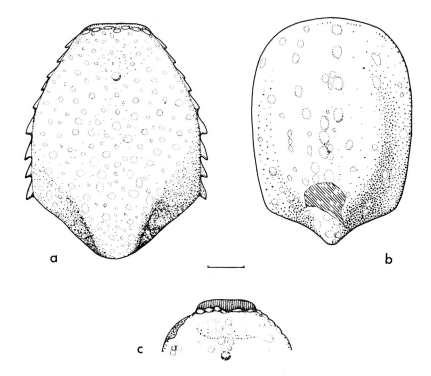

Fig. 4 *Ctenaspis russelli* sp. nov. Peel Sound Formation, Prince of Wales Island, N.W.T.
Scale = 10 mm.
    a. Dorsal shield (type: NMC 21702).
    b. Ventral shield (NMC 21703).
    c. Detail of rostral area. (NMC 218281).

Fig. 5 *Ctenaspis russelli* sp. nov. Peel Sound Formation, Prince of Wales Island, N.W.T.
Scales: a–d natural size; e twice natural size.
    a. Dorsal shield. NMC 21702.
    b. Ventral shield. NMC 21703.
    c. Detail of rostral area, showing large pineal node, large tubercles at rostral margin
       and oral margin. NMC 2128.
    d. The dorsal shield. NMC 21828.
    e. Detail of surface ornamentation of right lateral margin of ventral shield. NMC 21782.

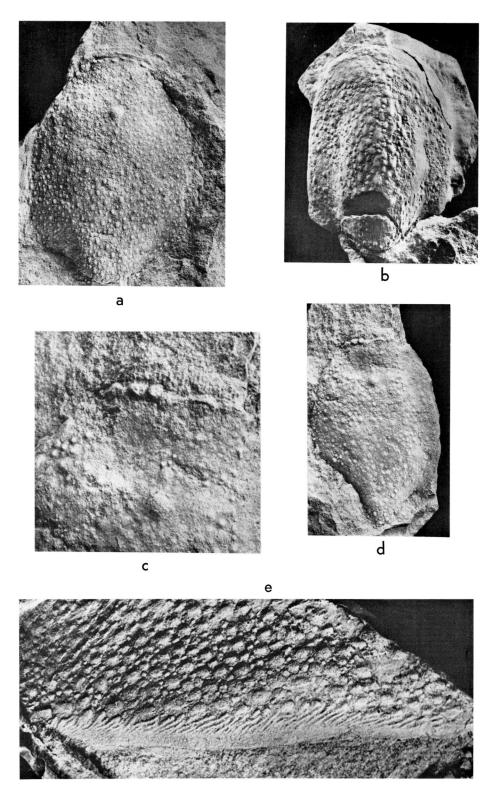

0.27). The lateral margins extend back from the orbits to the broadest part of the shield at about half its length and then converge only slightly towards the branchio-cornual angle. This margin is very coarsely serrate with 8–10 "cusps" projecting laterally. Beneath this margin the ventro-lateral lamina projects downwards and inwards and is about 2.5 mm deep. The posterior margin has lateral portions that broadly curve in towards a straight median part.

The ornamentation consists of large and flat smooth and round tubercles up to 2.5 mm in diameter with a row of elongate tubercles at the edge of the rostral margin, extending to the orbits. There is no median crest.

A number of specimens of this species reveal a feature which is not observed in other material and which is apparently the impression of a pre-oral surface or margin beneath the maxillary brim. The feature is only shown where the original bone of the shield has been broken away (see Figs. 4, 5). It is about 14 mm from side to side and 2.5 mm across, thus forming a narrow flat projecting shelf or margin above the transverse mouth. These impressions are closely striated from front to rear which suggests that in life this margin bore narrow ridges and grooves. No other cyathaspid has, to the writer's knowledge, a similarly grooved upper margin to the mouth, though presumably the pre-oral field in other genera may be homologous. The exact function of this margin is uncertain but it would no doubt have been adapted to the feeding mechanism and behaviour of the animal.

The ventral shield has a rounded anterior margin and sub-parallel straight sides which emphasize the branchio-cornual angle with the posterior margin. This kind of margin is similar to that in the dorsal shield but subtends a rather sharper V at the median line. The shield is deep and arched, but flatter towards the branchio-cornual angles: lateral laminae project vertically and uniformly for about 2.5 mm. The ornamentation appears to be somewhat coarser than on the dorsal shield, and the largest tubercles are situated near the mid-line, where they are ovate or elongate in outline and invariably smooth, as though worn. The ventro-lateral lamina has ornamentation of small, obliquely set, elongate tubercles or ridges.

No trace of the lateral line system has been detected nor have any scales been found.

Discussion

This is by far the largest *Ctenaspis* so far described, being about twice the size of *C. kiaeri* (Denison, 1964). It has a distinctive outline, especially in the ventral disc, while the dorsal disc is also distinguished from other species by the particular dentate or serrate character of its lateral margins. The large size of the pineal is remarkable as also is the extreme thinness of the shields.

The line of commissure of the two shields is comparatively straight and the branchial openings are directed posteriorly rather than laterally. Beneath the maxillary brim of the above-mentioned dorsal shield there is a wide slit separating the two shields, but the nature of the body wall and covering between orbits, maxillary brim and the anterior margin of the ventral shield is uncertain. It may be supposed that the mouth, extending between the pre-orbital processes was wide and immediately below the maxillary brim. If this were so, the mouth would be at a height of 8–10 mm above the deepest part of the body as in many, though not all, other cyathaspidids.

Ctenaspis ornata **sp. nov.**
Figs. 6, 7

Holotype
NMC 21856. Dorsal shield, incomplete. The species is so named because of the ornate surface markings of the shields and the presence of an ornamented dorsal spine.

Occurrence
Early Devonian (Gedinnian–?Siegenian). Sandstone facies of the Peel Sound Formation. Northeastern part of Prince of Wales Island, N.W.T. Found in loose scree on lower slope of Mt. Matthias.

Referred Material
NMC 21857–21859, present with the type on a single loose slab (Fig. 7d).

Diagnosis
A *Ctenaspis* with dorsal shield 4.5 to 5 cm long and with a width ratio of about 0.7, lateral brims serrate in the mid-third or anterior part, posterior margin smooth or with large tubercles, and a prominent short median spine. Ornamentation of large and small tubercles, transverse in rostral area. Ventral shield strongly arched, widest at posterior, margins smooth.

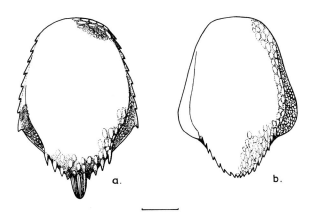

Fig. 6 *Ctenaspis ornata* sp. nov. Peel Sound Formation, Prince of Wales Island, N.W.T.
Restoration of fragmentary types. Scale = 10 mm.
a. Dorsal shield; only part of the ornamentation is shown.
b. Ventral shield; only part of the ornamentation is shown.

Description

The dorsal shield has a rather spade-shaped outline, widest at the (postero-lateral) cornual angles. The brim is generally smooth but with prominent tubercles in the middle third, producing a rather serrate appearance. The posterior margin tapers sharply towards the median line where a stubby spine is present. The spine projects about 3 mm beyond the brim and is longitudinally ridged. Several large elongate tubercles aligned antero-posteriorly mark the posterior brim on each side of the spine. The surface ornament of rounded to ovate tubercles is coarsest in the posterior part of the disc and finest near the lateral margins; at the edge of the disc runs a narrow smooth ridge. Close to the rostral margin the tubercles are elongate and transverse.

The ventral shield is similarly widest at the cornual angles. Its margins are smooth except posteriorly where well-spaced and prominent tubercles give it a rather serrate appearance. Although the material is incomplete it is clear that the ventral disc is strongly arched with its greatest depth near the posterior end.

Remarks

This species is poorly known from incomplete material, yet is sufficiently distinctive to be described as a separate species. Although the outlines of the two dorsal shields are somewhat different in detail they are similar enough to be regarded as belonging to the one species. It is, moreover, the only species yet known to possess a dorsal "spine". This spine is a low and robust outgrowth of the disc rather than a separate unit, some 6 mm long and projecting about 3 mm beyond the posterior margin. It has three stout and two minor longitudinal ribs.

The dorsal and ventral shields appear to fit together along a relatively straight line of commissure. The orbits were very small and the branchial openings prob-ably occurred as slits between the posterior margins of the two shields and the scaled trunk, as Denison found in *C. dentata* (1964, p. 441). The mouth parts are unknown as is the pineal area. No trace of scales has been found, although the slab containing these fossils includes many other cyathaspidid scales. The scaled body and tail were probably very similar to that in *C. obruchevi* (Fig. 3).

Fig. 7 *Ctenaspis ornata* sp. nov. Peel Sound Formation, Prince of Wales Island, N.W.T. Scales: a–c, approximately natural size; d, one quarter natural size.
   a. Incomplete dorsal shield showing spine at posterior margin. NMC 21856 (type).
   b. Incomplete ventral shield. NMC 21857.
   c. Incomplete dorsal shield showing serrate lateral margin and coarse tubercles of rostral area. NMC 21858.
   d. Slab of fine red sandstone with abundant shields of a small poraspidinid and the types of *Ctenaspis ornata* sp. nov. from Mt. Matthias.

a

b

c

d

Mode of Life

Like several other Heterostraci, *Ctenaspis* possesses a rather flat dorsal shield and a deeply moulded ventral shield. This would scarcely seem to be an adaptation for benthonic life, yet with its deep, laterally compressed body and powerful tail it may have been not unsuccessful among the benthos. *Ctenaspis* lacked paired fins and the caudal fin was apparently deep and symmetrical, forming a locomotory organ of considerable power. The belly was rounded, broad anteriorly but narrowing towards the hind end between the high cornual areas. The entire body was thus either rather fusiform or tadpole-like and can be visualised as remaining fairly stable while facing into a current. What, however, prevented it from toppling over sideways when at rest? The stout tail fin may have been directed to one side or the other as a prop. Although it lacked the large ventral lobe of, say, the pteraspids, it would have been sufficiently strong perhaps to prevent the body from tilting sideways. Very possibly, however, the animal spent much of its time with the ventral shield, trunk and tail at least partially buried in sediment (Fig. 8). Only the mouth, eyes and flat back would be level with the water-sediment interface. In this way the animal would be supported on an "even keel" and would, especially in a coarse

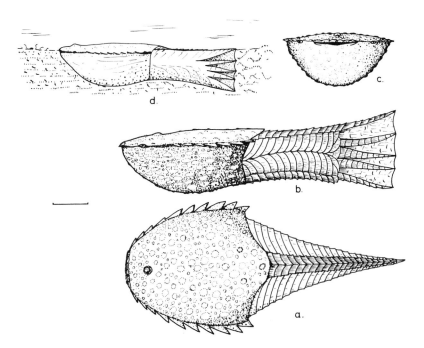

Fig. 8 *Ctenaspis*. Restoration of carapace and scaled body and tail. Scale = 10 mm.
    a. Dorsal aspect, about natural size.
    b. Lateral aspect about natural size.
    c. Anterior aspect about natural size.
    d. Possible resting position with greater part of body buried in sediment.

sandy stream bed, be well camouflaged by the rough flat surface of the dorsal shield. The narrow tranverse mouth would be advantageously placed to catch drifting food particles and benthonic organisms while the eyes would command a good field of vision. Small tentacles about the mouth might have existed but the pre-oral field may have been the site of cilia or mucus threads to which food adhered and was then drawn into the mouth.

Swimming and burrowing movements would be effectively conducted by vigorous movement of the tail. The fin web was strengthened by at least five "rays" and may have been flexible not only from side to side but from top to bottom. In passing, it may be noted that *Ctenaspis* offers another instance of a round-bellied heterostracan with a short rostrum possessing a relatively short diphycercal tail that was as deep as the body. Heterostraci with relatively long rostra (e.g. some pteraspids) appear to have had longer reversed heterocercal tails. (For discussion see Dineley and Broad, 1973.)

The coarse ornamentation and the serrate, or dentate, nature of the lateral margins on the dorsal shield may have been added to the stability of the animal in forming a highly irregular surface permitting movement of the animal forward over a granular bed, but inhibiting any tendency to slip backwards.

Phylogenetic Relationships

Ctenaspis exhibits an interesting array of features, and is seemingly primitive in the histology and architecture of the shield; there are no very obvious forms known from which it may have arisen. Kiaer (1930) thought that the ctenaspids were related to the Poraspidinae, and Denison (1964) felt that they could have derived from either Poraspidinae or Cyathaspidinae rather than independently from other cyathaspids, as Zych (1931) or Stensiö (1958) believed.

Stensiö (1964), however, suggested that the early ctenaspids gave rise to primitive forms with cyclomorial tesserae of which traces are visible in the tubercles of *Ctenaspis*. In the economy of structure, and the thin and seemingly primitive nature of the microstructure of the shields, *Ctenaspis* may indeed have been initial in such a line of evolution.

Denison (1964), in his survey of the Cyathaspididae erected the subfamily Ctenaspidinae to accommodate the single genus and four species known up to that date. I see no reason to differ from this view; however it should be noted that Obruchev (1964) accorded the ctenaspids the collective status of a family, Ctenaspididae. In this family were retained not only *Ctenaspis* but also *Bothriaspis* Obruchev (= *Ctenaspis kiaeri* Zych), *Cryptaspidisca* Strand (= *Allocryptaspis* Whitley), and possibly *Ariaspis* Denison (1963), *Putoranaspis* Obruchev (1964) and *Aphataspis* Obruchev (1964). It seems doubtful that the first two of these are closely related to *Ctenaspis*, and both are included by Denison in his subfamily Poraspidinae. Morphologically and in so far as the lateral line systems are concerned, they are widely different. *Ariaspis* is an undoubted cyathaspid. *Putoranaspis* and *Aphataspis*, however, are in gross outline and dimensions not dissimilar to *Ctenaspis*. Nevertheless their ornamentation is of fine, closely placed ridges rather than large discrete tubercles. In so far as I can judge, they must be related to *Ctenaspis* but not very closely, and they are regarded as members of the (new) families *Putoranaspididae* and *Aphataspididae* by Novitskaya (1971, 104–110). Halstead in a general review of the heterostraci (1973) retains the family Ctenaspididae in a new suborder Ctenaspida. He did not explain his division of the Order Cyathaspidiformes into Suborders, of which only the Hibernaspidida contains more than one family. They appear to be unnecessary taxa.

Novitskaya (1967, 1968, 1971) erected a possible phylogeny in which the 18 Siberian genera of amphiaspids are descended from a ctenaspid stock which probably was already distinct in the Silurian, but which remains to be discovered. Amphiaspids are now known in the lower (Pridoli) member of the Peel Sound Formation on Somerset Island (Broad, 1973). They appear to be primitive forms, large but rare and not occurring in later beds. By Gedinnian time the group was well established in Siberia but had perhaps disappeared from the Canadian faunas. The ctenaspids, on the other hand, continued to flourish in Canadian waters, and did not follow the path of the amphiaspid migration.

While Denison (1964) maintained that there are alternative interpretations of the shield of the early cyathaspid *Tolypelepis* to that offered by Stensiö (1958), it seems agreed that *Tolypelepis* is early and primitive, and has a dentine ornamentation found in no other cyathaspid. The ornamentation of *Ctenaspis* is also unique and might perhaps have been derived from a *Tolypelepis* type, the large flat tubercles corresponding to the central ridges of the "scale" components in the *Tolypelepis* shield. In the Ctenaspidinae the superficial layer of the shield has been lost and there has been no separation of the shield into epitega or distinct branchial plates. Perhaps the Ctenaspidinae arose from an early *Tolypelepis* stock or one of similar character. Fragments of *Tolypelepis* or genera closely resembling it are known from several early or mid-Palaeozoic formations in the Arctic Islands (Thorsteinsson, 1958).

Novitskaya (1971) sees a common ancestor to both the ctenaspids and the cardipeltids, and there are undeniable links between the cardipeltids and the psammosteids. Possibly several somewhat parallel lines of evolution lead from an early ctenaspid form, these being characterised by the amphiaspids, cardipeltids and psammosteids in Lower Devonian or Mid-Devonian rocks and taking place in separate provinces – the Siberian (Tungussian realm of Halstead and Turner, 1973), the Cordilleran and Caledonian respectively (Dineley, 1973).

Ctenaspids in the north of Canada during Silurian and Devonian times perhaps retained the basic primitive characteristics of the ctenaspid stock virtually unaltered. There was, however, an appreciable increase in the size of the animals indicating that their early adaptation to the local environments was highly successful and further changes were not forthcoming, at least during Peel Sound time.

Acknowledgments

The fossils were collected as part of the work of the Ottawa University Arctic Group to whom generous financial help was given by the Defence Research Board, Geological Survey of Canada, Department of Northern Affairs, National Museum of Canada and the Polar Continental Shelf Project. David Langley and Dr. D.S. Broad were invaluable and meticulous field assistants. The photographs were prepared by E. W. Seavill. To these institutes and friends the writer tenders his express thanks.

Literature Cited

BROAD, D.S.
    1973   Amphiaspidiformes (Heterostraci) from the Silurian of the Canadian Arctic Archipelago. Contr. Can. Paleont., Bull. Geol. Surv. Can., 222: 35–52.

BROAD, D.S., D.L. DINELEY AND A.D. MIALL
    1968   The Peel Sound Formation (Devonian) of Prince of Wales and adjacent islands: a preliminary report. Arctic, 21: 84–91.

BROWN, R.L., I.W.D. DALZIEL, AND B.R. RUST
    1969   The structure, metamorphism and development of the Boothia Arch, Arctic Canada. Can. J. Earth Sci. 6: 525–543.

DENISON, R.H.
    1963   New Silurian Heterostraci from southeastern Yukon (with Introduction by H.R. Hovdebo, A.C. Lenz and E.W. Bamber). Fieldiana: Geol., 14: 105–141.
    1964   The Cyathaspididae: A family of Silurian and Devonian jawless vertebrates. Fieldiana: Geol., 13: 309–473.

DINELEY, D.L.
    1965   Notes on the scientific results of the University of Ottawa Expedition to Somerset Island, 1964. Arctic, 18: 55–57.
    1966   Geological studies in Somerset Island, University of Ottawa Expedition 1965. Arctic, 19: 270–277.
    1973   The fortunes of the early vertebrates. Geology, 5: 2–20.

DINELEY, D.L. AND D.S. BROAD
    1973   *Torpedaspis*, a new genus of Cyathaspidid from the Peel Sound Formation, N.W.T. Contr. Can. Paleont., Bull. Geol. Surv. Can., 222: 53–92.

HALSTEAD, L.B.
    1973   The heterostracan fishes. Biol. Rev., 48: 279–332.

HALSTEAD, L.B. AND S. TURNER
    1973   Silurian and Devonian ostracoderms. *In* A. Hallam ed., Atlas of palaeobiogeography: 67–79. Amsterdam, Elsevier.

KERR, J.W. AND R.L. CHRISTIE
    1965   Tectonic history of Boothia Uplift and Cornwallis Fold Belt, Arctic Canada. Bull. Am. Assoc. Petrol. Geol., 49(7): 905–926.

KIAER, J.
    1930   *Ctenaspis*, a new genus of cyathaspidian fishes. Skr. Svalbard Ishavet, 33: 1–7.

1932   (ed. A. Heintz) The Downtonian and Devonian vertebrates of Spitsbergen. IV. Suborder Cyathaspida. Skr. Svalbard Ishavet, 52: 7–26.

MIALL, A.D.
1970a   Continental marine transition in the Devonian of Prince of Wales Island, Northwest Territories. Can. J. Earth Sci., 7: 125–144.
1970b   Devonian alluvial fans, Prince of Wales Island, Arctic Canada. J. Sed. Petrol., 40: 556–571.

MILES, R.S.
1973   An actinolepid arthrodire from the Lower Devonian Peel Sound Formation, Prince of Wales Island. Palaeontographica, Abt. A, 143: 109–118.

NOVITSKAYA, L.
1967   On the origin and certain lines of evolution of the amphiaspids (Heterostraci). Rep. Acad. Sci. U.S.S.R. 172: 1222–1225 (in Russian).
1968   New amphiaspids (Heterostraci) from the Lower Devonian of Siberia and classification of the amphiaspids. *In* Phylogeny and systematics of fossil fishes and Agnatha (in Russian): 45–62.
1971   Les amphiaspides (Heterostraci) du Dévonian de la Sibérie. Cahiers de paléont.: 1–130.

OBRUCHEV, D.V.
1964   Subclass Heterostraci (Pteraspides). *In* Fundamentals of palaeontology. Agnathans, Fishes. Ed. "Nauka", Moscow: 45–82 (in Russian).

STENSIÖ, E.A.
1958   Les cyclostomes fossiles ou ostracodermes. *In* P. Grassé, ed., Traité de zoologie, 13, 1: 173–425. Paris, Masson.
1964   Les cyclostomes fossiles ou ostrocodermes. *In* J. Piveteau, ed., Traité de paléontologie, 4, 1: 96–382. Paris, Masson.

THORSTEINSSON, R.
1958   Cornwallis and Little Cornwallis Islands, District of Franklin, Northwest Territories. Mem. Geol. Surv. Can., 294: 1–134.

THORSTEINSSON, R. AND E.T. TOZER
1963   Geology of northern Prince of Wales Island and northwestern Somerset Island. *In* Fortier, Y.O. *et al.*, Geology of the north-central part of the arctic archipelago, Northwest Territories (Operation Franklin). Mem. Geol. Surv. Can., 320: 117–129.

WILLS, L.J.
1935   Rare and new ostracodern fishes from the Downtonian of Shropshire. Trans. Roy. Soc. Edinburgh, 58, 2(18): 1934–1935.

ZYCH, W.
1931   Fauna ryb Dewonu i Downtonu Podola. Pteraspidomorphi: Heterostraci. Część IA: 1–91 (in Ukrainian).

A New Brachyopoid Amphibian

Everett C. Olson
University of California, Los Angeles

George E. Lammers
University of Winnipeg, Manitoba

Abstract

A siltstone nodule, provenance unknown, contains the skull and skeleton of a new temnospondylous amphibian. The age is presumably Permian. The specimen is designated as the type of a new genus and species, *Kourerpeton bradyi*. Its morphology is more or less intermediate between that of the Trimerorhachoidea and Brachyopidae. The presence of tightly joined pterygoid and parasphenoid bones is the principal basis for assignment of the new genus and species to the superfamily Brachiopoidea. The primitive position of the genus within the superfamily is the basis for erecting a new family, Kourerpetontidae, to accommodate it.

Introduction

The specimen described in this paper has a long and uncertain history as discussed in the following section. Consequently it has not been possible to obtain information upon where or in what beds it was found. In spite of this lack of information, the specimen provides important data on the evolution of brachyopoid amphibians, and so we have undertaken a description and discussion of its morphology and probable relationships to other amphibians.

The specimen is also of interest because of its mode of preservation. The bone was originally encased in a fine, siltstone nodule, superficially serpentine in appearance, and this undoubtedly accounted for its preservation through the vicissitudes it suffered after discovery. The bone was severely damaged by coarse crystallization of calcite and subsequently largely removed by the action of ground water. In many parts of the nodule this resulted in development of an excellent mould. The bones of the dermal surface of the skull were replaced, after removal, by siltstone, accurately enough that the dermal sculpture and some of the sutures were retained.

Where possible, before preparation, calcite was removed from the cavities by acid, and resin was injected to form casts of the bones. Parts of the palate, occiput, vertebrae, limbs and girdles yielded good casts. The resin appears as the black material in the plate. Elsewhere, after the nodule was split to reveal the moulds, latex casts were made with good results.

The preparation was carried out by Rick Lassen, who also did the photography. Mrs. Kathryn B. Bezy prepared the drawings. The work was supported by National Science Foundation Grant GB 31293x.

History of the Specimen

Specimen v-600 of the Manitoba Museum of Man and Nature has an interesting history, much of the earlier part now lost because of being passed on from owner to owner. The specimen was rediscovered by Major Lionel F. Brady, reportedly in the window of a barber shop in Bisbee, Arizona, where it was used as a decoration. Alternatively, the barber shop has been reported to have been in Mesa, Arizona. The barber believed the specimen had been collected from near Glen Rose, Texas. The exposed formations in the vicinity of Glen Rose are much younger than this specimen must be, so that this lead to the provenance is probably in error.

The specimen was catalogued and remained in the collection of the University of Arizona for many years where it had a long history of being subjected to casual scrutiny by visiting vertebrate palaeontologists. In 1967 it was withdrawn from the University of Arizona collection because of lack of provenance.

The junior author, upon taking a new position with the Manitoba Museum of Man and Nature in Winnipeg, brought the specimen to the attention of the senior author who acknowledged that it represented a significant stage of evolution and that an account should be published. It would be most desirable to collect additional specimens of this amphibian; any information on its possible provenance would be greatly appreciated.

Systematics

Class Amphibia
Subclass Labyrinthodontia
Order Temnospondyli
Superfamily Brachyopoidea
Family Kourerpetidae fam. nov.

Kourerpeton gen. nov.

Etymology
From *koura*, Greek for a barber's shop or place for the exchange of news and *herpeton*, Greek for reptile.

Diagnosis

A moderate-size amphibian, overall length about 60 cm. Pterygoid and parasphenoid rigidly joined; pterygoid descending in region of adductor fossa to produce a broadly arched palate posteriorly. Quadrate ramus of the pterygoid deep and extended somewhat posteriorly to the quadrate. Quadrate extending well below the level of the occipital condyle. Squamosal and quadratojugal slightly recurved around the posterior margin of the quadrate. Intertemporal bone absent. Exoccipital large and encompassing the vagal foramen; supraoccipital and otic elements not ossified. Branchial arches present and well ossified.

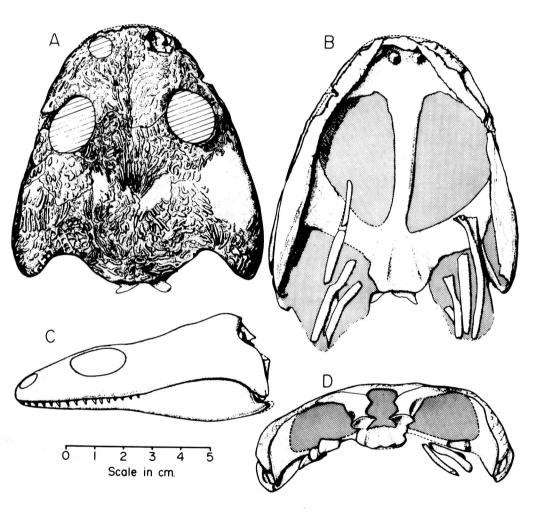

Fig. 1 Skull of *Kourerpeton bradyi* gen. et sp. nov.
  A. Dorsal.
  B. Ventral, slightly reconstructed by "removal" of matrix.
  C. Left lateral.
  D. Occipital, reconstructed by "removal" of matrix around branchial arches. Dotted lines represent areas obscured from view in this orientation by matrix.

Skull small relative to postcranium. Vertebrae typically rhachitomous with well-formed pleurocentra and intercentra. Humerus and femur small relative to body length. Measurements as in Table 1.

Kourerpeton bradyi sp. nov.
Figs. 1, 2, 3, 4

Diagnosis
As for genus.

Holotype
Manitoba Museum of Man and Nature, MMMN-V600. A skull and skeleton preserved largely as a mould in a siltstone nodule. The specific name is after the "rescuer," Major Brady.

Horizon and locality
Unknown. See introduction and discussion in preceding section. Probably from the Permian, but not from any site with which the writers are familiar.

Description
Very generally the skull resembles that of various trimerorhachoids and brachyopoids in outline and proportions. The postcranium is comparable to that of various trimerorhachoids and has no features which exclude it from the superfamily. The postcranium of brachyopoids is little known, so that few comparisons can be made.

SKULL (Figs 1A–D, 2A–B)

1. *Proportions*: Dimensions and proportions are entered in Table 1 and shown in Fig. 4. Comparative ratios between various measurements and skull lengths for *Kourerpeton* and other genera are given in Table 1.

The width to length ratio, 1.09, is comparable to that in *Trimerorhachis insignis* and *T. rogersi* and is in the low end of the range of brachyopids. Among the trimerorhachoids, *Isodectes* and *Dvinosaurus* have relatively broad skulls. The generic name *Isodectes* is used throughout, following Donald Baird as cited in Welles and Estes (1969, p. 29), for *Eobrachyops* Watson. The measurements are those for *Eobrachyops townsendi* Watson as given by Welles and Estes. The width of the skull in the orbital region of *Kourerpeton* is relatively greater than that in *Trimerorhachis insignis* and *Isodectes* and more or less comparable to that of *Dvinosaurus* and *Batrachosuchus*. This is shown in the measurements in the relationships of both interorbital width and orbital width to skull length.

The skull is moderately deep posteriorly as shown in Figure 1c. The ratio of greatest skull depth to length is about 0.34 in *Kourerpeton* as compared to a mean ratio of 0.23 for three skulls of *Trimerorhachis rogersi* (Olson, 1955), a deep-skulled species. Skulls of the other genera used in the comparisons are as deep or deeper.

2. *Dorsal aspect*: Figs. 1A, 2A. Because the dermal bones were replaced by matrix many of the sutures have been obscured. The dermal reticulation, however,

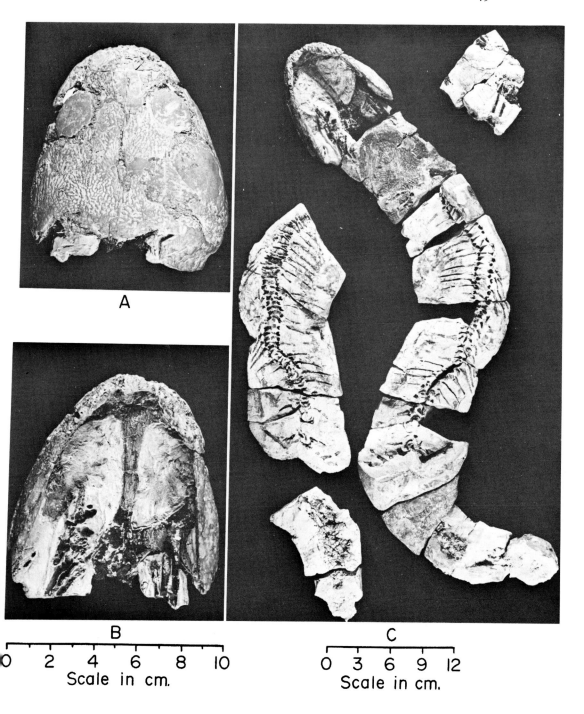

Fig. 2 *Kourerpeton bradyi* gen. et sp. nov.
A. Skull in dorsal aspect.
B. Skull in ventral aspect.
C. Skull and skeleton in ventral aspect as seen in split nodule.

is clear in places and aids in determination of bone limits. In a few places, as shown in the figure, sutures can be traced. One of these is the area of the postorbital region of the left orbit. As shown, the postorbital bone is very large and occupies the full area which includes both the postorbital and intertemporal in the trimerorhachoids. The postfrontal appears to be very small, producing the unusual condition in which the postorbital meets the parietal. This is similar to the condition in *Dvinosaurus* as described by Bystrow (1939), but he noted that the sculpture suggested the presence of an intertemporal. Following this, Romer (1947) also considered the apparent contact of the postorbital and parietal as evidence that an interparietal had intervened. The sculptural pattern on *Kourerpeton* gives no indication whatsoever of the presence of an intertemporal. If one were present, it had completely fused with the postorbital, with development of a perfectly matched sculptural pattern. The only reasonable interpretation from the evidence is that none was present.

Romer (1947), among others, suggested that the short, broad skull of *Dvinosaurus* resulted from neoteny, indicated by the presence of branchial bars. *Kourerpeton* is similarly perennibranchiate, but there are few other features that can be clearly interpreted as neotenic. Only the lack of ossification seen in occipital aspect may support such a conclusion. The specimen as a whole suggests a fully mature, aquatic amphibian.

An oddity relative to the presumed aquatic habitat is the absence of grooves for the sensory canal system of the head. Only by an unwarranted stretch of the imagination can possible courses of a canal system be discerned. This does not seem to be a matter of preservation as the external sculpture is well shown in areas where canals must have existed. Rather, it would seem that the canals were not impressed upon the bone, although they were probably present in the dermis.

The area of the otic notch is shown in dorsal view. Either the notch was nonexistent or it was represented by a very gentle flexure. In Figure 1A the exoccipitals can be seen to slope moderately to the posterior, but there is no evidence of the extensive projection of the occipital region back of the dorsal platform. The lack of ossification in the supraoccipital and otic elements (as well as damage to the occipital stem and condyle) may have been moderately extensive, but almost certainly not as extreme as in fully developed brachyopids.

3. *Occipital aspect*: Fig. 1D. The occipital condyle was destroyed at the junction of two parts of the nodule. The mould gives little detail of its structure. Exoccipitals are large and fully ossified, encompassing the vagal foramen and extending well laterally below the dermal surface of the skull. The supraoccipital was unossified as were the otic elements. It is possible that this is the result of neoteny, but the condition resembles that of brachyopids in which this is not a factor.

The left quadrate-articular region is shown in Fig. 1D, slightly reconstructed. The quadratojugal and squamosal carry around the quadrate posteriorly, but only slightly and do not produce the concave surface found in brachyopids. The broad quadrate ramus of the pterygoid is shown (see also Fig. 1C) and it projects somewhat posteriorly to the quadrate. Whether a groove exists between it and the quadratojugal-squamosal as in brachyopids is uncertain.

The branchial arches are shown in posterior aspect. Preserved as moulds filled with resin, these are partly embedded in matrix in the specimen (Fig. 2B), and in

Figure 1D are reconstructed as if the matrix were not present. The dashed lines on the occipital aspect of the skull show the general areas covered by matrix.

4. *Lateral aspect*: Fig. 1C. Much of what is illustrated is self-explanatory and has been noted earlier. The posterior depth of the skull, the slightly raised orbital region, and the ventral extension of the quadrate are shown. The shape of the lower jaw, with the relatively ventral position of the articulation and the short retroarticular process are points of interest. The marginal teeth in the maxilla and premaxilla are reconstructed from alveoli which give some basis for spacing and width, but not for length. A few lower teeth can be seen on the inner side of the jaw. They are simple cones with lengths comparable to those shown for the anterior maxillary teeth in the drawing.

5. *Palatal view*: Figs. 1B, 2B. Fig. 1B has been but little reconstructed, with modification only at the anterior end of the palate and in the area of the branchial arches, where some matrix has been "removed". The branchial elements are as shown. The most lateral one on the right side as viewed is best preserved and shows a strong vertical extension anteriorly and a termination that suggests continuation in cartilage. Four elements appear to be present on each side and these presumably represent ceratobranchials 1 to 4, as in *Dvinosaurus*. No evidence of any other branchial elements has been found.

Anteriorly a moderately large tusk is present on each vomer. As preserved each is encased in matrix and damaged by recrystallization. The identification of the bones on which they occur as vomers is based purely upon position. Additional palatal tusks are present. One occurs posterolaterally to the choana and two, possibly three, occur more posteriorly, probably on the palatine.

An important structure of the palate is the strong fusion, or possibly sutural junction, between the parasphenoid and the pterygoid in the basipterygoid region. This contrasts sharply with the condition in trimerorhachoids and is like that in brachyopids. Anteriorly the parasphenoidal rostrum expands somewhat and does not appear to pass under (dorsal to) the vomers as it tends to do in brachyopids.

POSTCRANIUM (Figs. 2C, 3)

1. *Vertebrae and ribs*: Fig. 3A,B. From the occiput to the level of the femur (no pelvis has been identified) there are about 40 vertebrae. In addition about 18 cm of the caudal region are present, although preservation is very poor and individual vertebrae cannot be seen.

In the presacral column the vertebrae are typically rhachitomous. Both pleurocentra and hypocentra are well ossified. The two halves of the former nearly meet ventrally at the midline. Dorsally they taper to slender processes that almost close the gap above the notochordal tube. Each hypocentrum carries a small facet for articulation of the capitulum of the rib. Hypocentra are strong and much like those of *Neldasaurus* (Chase, 1965). They too have well-formed dorsal rami. As far as can be told the vertebrae differ little along the column. Those shown in the figure are based on vertebrae from the anterior dorsal part of column where it was possible to get good resin injections. Latex casts from other areas reveal similar characteristics. The anteriormost vertebrae cannot be seen because they are encased in matrix of the part of the nodule which carries the shoulder girdle, which we did not wish to damage.

Ribs are short, rather broad and little curved. The heads are moderately broad

with the tuberculum and capitulum little differentiated. The ribs extend along the
full presacral column diminishing somewhat in size posteriorly.

2. *Girdles*: Figs. 1D, 2C. The shoulder girdle is fairly well preserved, showing both
dorsal and ventral surfaces, but there is no evidence of a pelvic girdle, which originally may have been in the area posterior to the femur where preservation is poor.

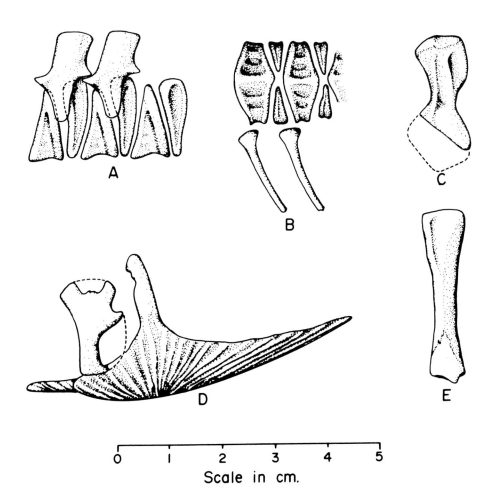

Fig. 3 Postcranial bones of *Kourerpeton bradyi* gen. et sp. nov.
   A. Anterior vertebrae in left, lateral aspect, reconstructed from resin and latex casts.
   B. Anterior vertebrae in ventral aspect showing also nature of ribs, reconstructed from
      latex cast.
   C. Humerus, ventral aspect, reconstructed from resin cast with dotted line representing
      missing portions.
   D. Shoulder girdle in right lateral aspect. Reconstructed from mould and from resin
      casts.
   E. Femur, ventral aspect reconstructed from resin cast. For these elements also see
      Fig. 2C.

The pectoral girdle has a broad, somewhat diamond-shaped interclavicle and strong clavicles. The ascending ramus of the clavicle is preserved as a mould and was cast in resin. The scapula was similarly preserved, although ossification does not appear to have been complete. The reconstruction on Figure 2D is based on the ventral impressions and the more dorsal casts. The girdle shows no special features and is similar to that in many temnospondyls.

3. *Limbs*: Figs. 2C, 3C,E. Only a partial humerus and a femur are present. The more distal elements extended beyond the limits of the nodule and were, it would appear, never preserved. The humerus is small, with an estimated length of 25 mm. Its length, compared to the basal width across the clavicles, is about 40% of the latter. In *Neldasaurus* the same comparison yields a value of about 70% and this is relatively low for trimerorhachoids. Considerable reduction of limbs from the primitive condition appears to have occurred.

The femur is relatively long and slender. The ends are somewhat expanded and the shaft is narrow, about 4.5 mm at the narrowest part. The length is about 30 mm. A fairly strong trochanter is present on the ventral surface a short distance from the medial end of the bone. Both the femur and humerus show no notable differences from comparable elements in the trimerorhachoids.

Table 1. **Measurements in mm of the skull of *Kourerpeton bradyi* (measurements and abbreviations as in Figure 4) and ratios of various measurements of the skull relative to skull length for *Kourerpeton bradyi* and selected trimerorhachids and a brachyopid. [a]Based on *Eobrachyops townsendi* of Watson. *2 Specimens. **Range within genus. ***Estimates from figures.**

Measurements (mm)	*Kourerpeton bradyi*
Sk_1	82
Sk_w	89
$O\text{-}S_1$	20
$O\text{-}Pi_1$	10
O_1	21
O_w	18
IO_w	22
INA_w	14

Proportions	*Kourerpeton bradyi*	*Trimerorhachis insignis*	*Isodectes*[a] *megalops*	*Dvinosaurus primus*	*Batrachosuchus* (sev. sp.)
$Sk_w:Sk_1$	1.09	1.07 $\bar{x}$ of 6	1.20	1.17* 1.24	1.19–1.50**
IO_wSk_1	0.26	0.17 $\bar{x}$ of 8	0.34	0.33* 0.41	0.43–0.50**
$O\text{-}S_1:Sk_1$	0.20	0.28 $\bar{x}$ of 8	0.29	0.29* 0.31	0.19–0.25**
$INA_w:Sk_1$	0.17	0.26 $\bar{x}$ of 4	0.48	—	—
$O_1:Sk_1$	0.25	0.20 $\bar{x}$ of 6	0.22***	0.28* 0.27	0.23–0.30**
$O_w:Sk_1$	0.22	0.18 $\bar{x}$ of 7	0.15***	0.23***	0.18***

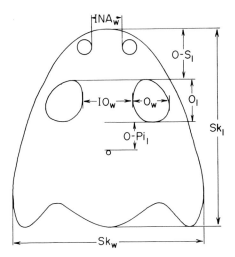

Fig. 4 *Kourerpeton bradyi* gen. et sp. nov. Diagram of skull in outline showing measurements used in Table 1. Abbreviations: INA_w, internarial width; IO_w, interorbital width; O_l, orbital length; O_w, orbital width; $O-Pi_l$; orbital – pineal length; Sk_l, skull length from tip of snout to posterior margin of skull platform; Sk_w, greatest width of skull.

Relationships

The relationships indicated in the section on systematics show *Kourerpeton* as the member of a new family, *Kourerpetidae*, within the superfamily Brachyopoidea. According to usual assignment of this superfamily, this technically places the new family under the suborder Stereospondyli. The vertebral column, however, is not stereospondylous. It appears that in some of the brachyopids this was also the case. The various genera appear to fit well into the informal grade of neorhachitomes, one between rhachitomous and strictly stereospondylous vertebral conditions. Within this grade considerable variation in the condition of the pleurocentrum and hypocentrum exists.

Within the temnospondylous amphibians, affinities of *Kourerpeton* are clearly with the Trimerorhachoidea on the one hand and the Brachyopoidea on the other. The former includes three families (Chase, 1965); Saurerpetontidae, Trimerorhachidae and Dvinosauridae. With reference to these, *Kourerpeton* has some of the distinctive structures found in the first and some of those found in the Dvinosauridae. The Brachyopoidea, following Romer (1966), includes the single family Brachyopidae, and it is with this that comparisons can profitably be made.

At one time or another a rather large number of temnospondyls have been considered as brachyopids or as ancestral to the family. In recent studies by Chase (1965), and by Welles and Estes (1965, including citations of unpublished work by Donald Baird) and Shishkin (1966), most of these have been removed from the family and their ancestral position has been challenged. These studies are but the most recent among many that have considered evolutionary relationships and generally challenged efforts to bridge the gap between trimerorhachoids and brachyopids.

Although some genera have paralleled brachyopids in a few features, we are currently faced with the lack of evidence of any generally acknowledged ancestor of this group. *Isodectes megalops* (Cope) and *Dvinosaurus primus* Amalitsky have received most attention.

It is now generally agreed, following Romer (1947), that *Dvinosaurus* is a persistent member of the Trimerorhachoidea, the representative of the monogeneric family Dvinosauridae. *Isodectes* was treated at length as a brachyopid, under the name of *Eobrachyops*, by Watson (1956). It was considered to be a trimerorhachoid by Chase (1965), also under the name of *Eobrachyops*. The latter placement is reasonable, but the genus does have a number of features that suggest a trend in the direction followed by the brachyopids. Among these are the large exoccipitals which include the vagal foramen, a reduced otic notch, and an extension of the quadratojugal and squamosal around the posterior flank of the quadrate. These are "advanced" characters, but as Chase noted there are also specializations – a small tabular, reduced sensory canals, small maxillae – which bar it from direct ancestry. Along with *Saurerpeton* and *Acroplous*, it belongs in the family Saurerpetontidae within the Trimerorhachoidea, a family that is characterized by various brachyopid-like features.

Chase (1965) and Welles and Estes (1969) published lists of characters which defined the Trimerorhachoidea and Brachyopidae respectively. The lists serve their purposes broadly within the Labyrinthodontia, but because of the many common characters between the two groups they contain many overlaps. From these lists, plus some items contained only by one or the other, features which differentiate the two groups may be sorted out as follows:

	Character	Trimerorhachoidea	Brachyopidae
1.	Basipterygoid joint	Open and freely movable	Parasphenoid and pterygoid tightly joined by fusion or suture
2.	Intertemporal bone	Present	Absent
3.	Occipital condyle	Single, except in *Dvinosaurus*	Double
4.	Quadratojugal-squamosal	Not encompassing quadrate posteriorly; some trend in this direction in Saurerpetontidae	Encompassing quadrate posteriorly, producing concave surface; groove separating from pterygoid
5.	Quadrate	At or near level of occipital condyle in horizontal plane; *Isodectes* excepted	Well below level of occipital condyle in horizontal plane
6.	Retroarticular process	Short	Long
7.	Pterygoid	Quadrate ramus relatively narrow dorso-ventrally, little reflected downward medial to adductor fossa; not extended posterior to quadrate; *Isodectes* partial exception	Quadrate ramus deep and reflected downward around adductor fossa; extended posterior to quadrate
8.	Palatal configuration	Little arched posteriorly, *Isodectes* exception	Arched to form gentle, inverted U posteriorly

Kourerpeton conforms to brachyopids in points 1, 2, 5, 7 and 8. It conforms to trimerorhachoids in point 6. It shows only moderate development of point 4; point 3 is unknown. On the basis of skull and jaw features, thus, *Kourerpeton* appears structurally somewhat intermediate between trimerorhachoids and brachyopids. The preponderance of characters indicates a greater proximity to brachyopids than to any family of the trimerorhachoids. If *Kourerpeton* arose within the trimerorhachoids then the probable source is the family Saurerpetontidae. Nothing that is now known of the skull and jaws excludes it from a position ancestral to the brachyopids.

Even though the genus was most likely perennibranchiate, the evidence that it was neotenic seems slight, as discussed earlier. As an adult amphibian *Kourerpeton* occupies an appropriate position as an ancestor of the brachyopids. Whether or not any of the latter were perennibranchiate is uncertain, although there is no positive evidence that this was the case. This feature in *Kourerpeton* may thus be a specialization not to be expected in brachyopid ancestry.

The postcranium is distinctly rhachitomous, whereas brachyopids are usually considered as stereospondyls. This assignment as far as vertebrae are concerned rests on slender evidence, for vertebrae are not well known. Intercentra associated with *Batrachosaurus*, described by Wells and Estes (1969), are of the rhachitomous type. No pleurocentra have been found, however. The anteriormost vertebrae are like those of other stereospondyls, as is to be expected in view of the broadly double occipital condyle. Although brachyopids apparently fall in the neorhachitomous grade, there is no evidence that any had as distinctly a rhachitomous condition such as is found in *Kourerpeton*.

The best estimate of the phylogenetic position of *Kourerpeton* places it between the trimerorhachoids and brachyopids. The skull has developed more brachyopid features than has the postcranium. In an evolutionary sense the animal seems to provide an excellent link between the two. Taxonomically it poses some problems. The genus cannot be placed among the Trimerorhachoidea without doing serious violence to the definition of this group. Most important is the tightly joined parasphenoid and pterygoid. It cannot be placed within the family Brachyopidae without a serious modification of the definition of this family, which seems undesirable. Of the remaining alternatives – of leaving it unassigned or erecting a new family – we have chosen the latter as more satisfactory. It is on this basis that the family Kourerpetontidae has been established. The family is placed within the superfamily Brachyopoidea because the preponderance of characters of *Kourerpeton* are more like those of the Brachyopidae than of any other known family of temnospondyls.

Literature Cited

BYSTROW, A.P.
   1938   *Dvinosaurus* als neotenische Form der Stegocephalen. Acta. Zool., 19: 209–295.

CHASE, J.
   1965   *Neldasaurus wrightae*, a new rhachitomos labyrinthodont from the Texas Lower Permian. Bull. Mus. Comp. Zool., 133: 152–225.

OLSON, E.C.

1955   Fauna of the Vale and Choza: 10, *Trimerorhachis*, including a revision of pre-Vale
        species. Fieldiana: Geology, 10: 225–312.

ROMER, A.S.

1947   A review of the Labyrinthodontia. Bull. Mus. Comp. Zool., 99: 3–352.
1966   Vertebrate Paleontology, 3rd ed. Chicago, 468 pp.

SHISHKIN, M.A.

1966   A brachyopid labyrinthodont from the Triassic of the Russian Platform. Paleont.
        Jour., 1966, no. 2: 93–108 (in Russian).

WATSON, D.M.S.

1956   The brachyopid labyrinthodonts. Bull. Brit. Mus. (Nat. Hist.): Geology, 2:
        315–392.

WELLES, S. AND R. ESTES

1969   *Hadrokkosaurus bradyi* from the Upper Moenkopi Formation of Arizona. Univ.
        Calif. Publ. Geol. Sci., 84: 1–56.

Eosuchians and the Origin of Archosaurs

Robert L. Carroll
Curator of Vertebrate Paleontology, Redpath Museum,
McGill University, Montreal

Abstract

An eosuchian reptile from the *Cistecephalus* zone (Upper Permian) of South Africa, *Heleosaurus scholtzi* Broom, has a marginal dentition nearly identical to that of the archosaur *Euparkeria capensis* Broom, dermal armour-plates along the vertebral column, and a femur indicative of an upright posture. This species, which is generally similar to *Youngina* in morphology, demonstrates that archosaurs could have evolved from eosuchians. The eosuchian-archosaur transition probably took place within a lineage of small, active, terrestrial carnivores. *Euparkeria* probably best characterizes primitive thecodonts, while *Proterosuchus-Chasmatosaurus* and *Erythrosuchus* represent early specializations toward larger body size and semiaquatic habits. Eosuchians may have evolved from such forms as the late Pennsylvanian captorhinomorph *Petrolacosaurus kansensis* which has a fully developed diapsid configuration of the skull.

Introduction

Among his various fields of interest within vertebrate paleontology, Dr. Russell has contributed substantially to our understanding of Late Mesozoic dinosaurs. Dinosaurs have in fact provided much of the stimulus for palaeontological research in Canada. Although the problems of the origin of this group and of other archosaurian assemblages lie somewhat outside the realm of his own research, discussion of this subject may nevertheless be an appropriate topic for this Festschrift.

The origin of the archosaurs has been subject to considerable speculation in recent years. Watson (1957) took what might be considered the "classical" view that all reptiles with a primitively diapsid skull configuration had a common ancestry—squamates, rhynchocephalians, and the ancestral archosaurs might all be traced to eosuchians such as *Youngina.* Romer (1956; 1966), in contrast, has held that archosaurs and lepidosaurs should be viewed as fundamentally distinct groups, although he considers that both ultimately arose from the captorhinomorphs. Romer (1967) cited a number of characteristics of the ear region and jaw suspension that appeared specifically to preclude eosuchians from the ancestry of archosaurs. A strikingly different suggestion as to the ancestry of the group was made by Reig (1967; 1970) who cited numerous similarities in skull and postcranial morphology between early archosaurs and varanopsid pelycosaurs. These similarities were discussed at some length by Romer (1971) who concluded that most could be attributed to common inheritance of primitive features and could not be used to indicate special relationships.

The most recent contribution to the topic was by Cruickshank (1972) with his description of *Proterosuchus-Chasmatosaurus,* one of the most primitive thecodonts. Cruickshank agreed with Romer in minimizing the significance of Reig's arguments for pelycosaur affinities. He went on to emphasize certain points of resemblance between *Proterosuchus* and rhynchosaurs that might be suggestive of close affinities, or in any case, close relationships between archosaurs and primitive lepidosaurs in general.

As a result of the descriptive work of Hughes (1963) and Cruickshank (1972) on *Proterosuchus* and Ewer (1965) on *Euparkeria* we have a fairly complete knowledge of the most primitive known Triassic archosaurs. The Russian genus *Archosaurus* (Tartarinov, 1961), although from the latest Permian, appears very similar to *Proterosuchus* and does little to increase our understanding of the ancestry of the group. In the earlier Permian beds, no genera have been recognized as more primitive archosaurs or as appropriate ancestral forms within more primitive taxonomic groups.

All previously described eosuchians are being studied by the author and Mr. C.E. Gow of the Bernard Price Institute in an effort to determine more precisely the relationship of these forms to both archosaurs and lepidosaurs. There are actually very few specimens that can definitely be included among the eosuchians, accepting *Youngina* as characteristic of the group. Of these, one contributes significantly to our understanding of the origin of archosaurs. The main purpose of this paper is to describe this specimen.

Methods and Materials

In 1907, Broom described a single specimen from the neighbourhood of Victoria West as the type of a new genus and species, *Heleosaurus scholtzi.* He published figures and a brief description, concluding that the specimen most closely resembled mesosaurs. Although no additional work was done on the specimen in the intervening years, Romer (1956) included the genus in the eosuchian family Younginiidae.

Much of the skeleton is preserved in counterpart blocks (Fig. 1). Two features that are at once evident (and were commented on by Broom) are the presence of dermal armour in the form of small plates scattered along the vertebral column, and sharp, blade-like teeth, bearing serrations on both the anterior and posterior margins. Both are characteristic of primitive carnivorous archosaurs, notably the genus *Euparkeria*. They provided impetus for a thorough study of the animal. The blocks had been split in such a manner as to divide the animal in a horizontal plane, the dorsal side in one block, the ventral in the other. Most of the vertebral column was split at the level of the zygapophyses. The bone was much softer than the matrix and in several places had rotted away, leaving only an impression. Orthodox preparation appeared difficult and impractical. After some experimentation in areas where the bone was largely missing, it was concluded that the most efficient way of preparing the specimen was to remove the bone and utilize the resulting casts in the matrix to prepare latex moulds. Because the matrix was both very hard and extremely fine-grained, very great detail of skeletal morphology could be disclosed by this method. In contrast with the many specimens prepared in a similar manner from the gas coal deposits of Linton, Ohio, and Nýřany, Czechoslovakia, the individual bones of this animal were preserved in nearly their original three-dimensional configuration, permitting lateral as well as dorsal and ventral illustrations to be made of the vertebrae and skull. Unfortunately, the skeleton had disintegrated to some extent prior to burial, with the loss of the dorsal surface of the skull, the forelimbs and the distal portion of the rear limbs. The tail was either lost at that time, or became separated from the remainder of the block subsequently.

Systematics

Class Reptilia
Order Eosuchia
Family Younginiidae

Genus *Heleosaurus* Broom, 1907

Diagnosis
Younginid reptile with thecodont dentition and dermal armour of thecodont pattern. Approximately twelve laterally compressed, recurved and sharply pointed teeth in maxilla. Teeth commonly bear minute serrations on posterior keel, and rarely on anterior edge. Jugal bears lateral process and posterior system of ridges and grooves. More than a single paired row of dermal ossicles along vertebral column, with approximately 2.5 ossicles along length of a single vertebra. Pectoral and pelvic girdles generally primitive. Femur similar to that of crocodilians, probably capable of vertical orientation.

Heleosaurus scholtzi **Broom, 1907**

Holotype

South African Museum, No. 1070 (Fig. 1). The only known specimen.

Type Locality

Victoria West, Cape Province, South Africa. Lower Beaufort, low in the *Cistecephalus* zone (Kitching, 1971 and pers. comm.).

Description

SKULL ROOF

As originally exposed, the counterpart blocks showed only the palate, in dorsal and ventral views, together with the lower jaws and the marginal bones of the cheek region. The dorsal block was itself split, and the matrix removed down to the level of the bone, but this failed to reveal any elements that were not visible ventrally.

Almost nothing of the skull roof is preserved. A small area of bone that may represent the nasal is visible behind the left vomer (Fig. 1). More important, a part of the right parietal can be seen between the overturned right palatine and the right quadrate. Judging from the configuration of the skull of *Euparkeria* and *Youngina*, it is the narrow portion of the bone between the upper temporal opening and the occipital surface. This is also the position that this fragment occupies relative to most of the remainder of the skull. This provides the only direct evidence that this particular skull had an upper temporal opening.

The ventral margins of both the lower temporal opening and the orbit are established by the preserved portions of the jugal and quadratojugal (Fig. 2A). Both openings were evidently quite large, and may be restored as similar to *Youngina*. The orbit was apparently much more nearly circular than that of *Euparkeria*. A groove may be seen at the posterodorsal margin of the quadratojugal, indicating that this bone lacked the dorsal process seen in archosaurs, and that the posterior margin of the lower temporal opening is formed solely by the squamosal. The anterior extension of the quadratojugal is loosely set in a depression on the lateral margin of the jugal. The two bones overlap one another for a considerable distance. The jugal is characterized by having a substantial lateral tuberosity on the ventral margin, just anterior to the level of the ascending process. More posteriorly, the margin of the bone is marked by longitudinal grooves and ridges. It is probable that these rugosities and the lateral process were associated with a ligament running back to the quadrate that assisted in stabilizing an otherwise fairly loose lower temporal bar.

In both early archosaurs and rhynchosaurs, as well as the ancestors of squamates, there was considerable mobility of the quadrate. Such was evidently the case in this form as well. A portion of the quadrate is preserved in the dorsal block. What is present suggests similarity with that of *Youngina* rather than with *Euparkeria*. The bone has only a limited lateral exposure and may have been fairly short. It has a vertical ridge that may have marked the position of attachment for a tympanum extending quite low on the cheek. A large quadrate foramen is evident on the lateral margin, above the articulating surface. Only the posterior portion of the articulating surface is visible. There are two prominent condyles extending forward and angling slightly medial to the long axis of the skull.

PALATE AND BRAINCASE

Although all the bones are somewhat disarticulated and the right palatine is turned completely over, almost the entire palate can be reconstructed (Fig. 2B). Only the posterior portion of the vomers, the anterior end of the maxillae, and the premaxillae are missing. The outline is roughly triangular, closely resembling that of romeriids, *Youngina*, and *Euparkeria* (Fig. 11 below). The occipital condyle lies just anterior to the back of the jaw articulation. The internal nares are long and narrow. Both vomers and palatine bones bear ridges on the margins of these openings. The interpterygoid vacuity is narrowly triangular. The subtemporal fenestrae are relatively small. There appear to be small palatine fenestrae, but neither palatine bone is sufficiently well preserved to show its extent. All of the paired palatal bones bear denticles, in the pattern common to captorhinomorphs and pelycosaurs. Seven teeth are borne on the margin of the transverse flange of each pterygoid. Two rows of denticles extend anteriorly from the basicranial articulation. The more lateral row extends onto the palatine. This continues anteriorly, to the medial margin of the internal naris. A few scattered denticles can be seen on the preserved surface of the pterygoid and palatine, as well as the ectopterygoid. The vomers have a row of denticles on their medial margin. Denticles are also seen on the vomers of *Euparkeria* (Gow, 1970).

The ectopterygoid is nearly square. The lateral margin is thickened where it abuts against the jugal. A short ventral ridge extends posteriorly where the ectopterygoid makes contact with the lateral margin of the pterygoid. The vomers are short, narrow bones, fitting posteriorly between the palatine and pterygoid. The palatine is a simple flat bone, without the ridges noted in millerettids (Gow, 1972). The lateral margin is not noticeably thickened where it was in contact with the maxilla. The configuration of the pterygoids is nearly identical with that seen in romeriids, but does not differ significantly from that in *Euparkeria* except that in that genus teeth are not present on the transverse flange of the pterygoid.

The parasphenoid-basisphenoid complex is complete, except for the anterior portion of the cultriform process. Its complexity compares well with that of *Proterosuchus* (Fig. 3). Unfortunately this area is not adequately preserved in ventral view in any of the specimens of *Euparkeria*. In contrast with romeriids, a sharp parabolic ridge divides the plate of the parasphenoid just posterior to the basicranial articulation. Grooves for the carotid arteries terminate in foramina near the anterior margin of the ridge. A ventral keel of the cultriform process terminates in a rugose ridge at the apex of the parabola. The relationship of the basicranial processes and the pterygoids remains essentially as in romeriids, and resembles closely that of *Proterosuchus* as well.

The basioccipital is disarticulated and lies alongside the axis vertebra. The ventral surface that was normally covered by the paraphenoid is marked by longitudinal grooves and ridges. At the posterior margin of this area, paired lateral processes extend dorsally toward the exoccipitals. The exoccipitals are not in place. A bone that may be so identified lies among the other occipital and cervical elements at the anterior end of the vertebral column. It is not well enough preserved

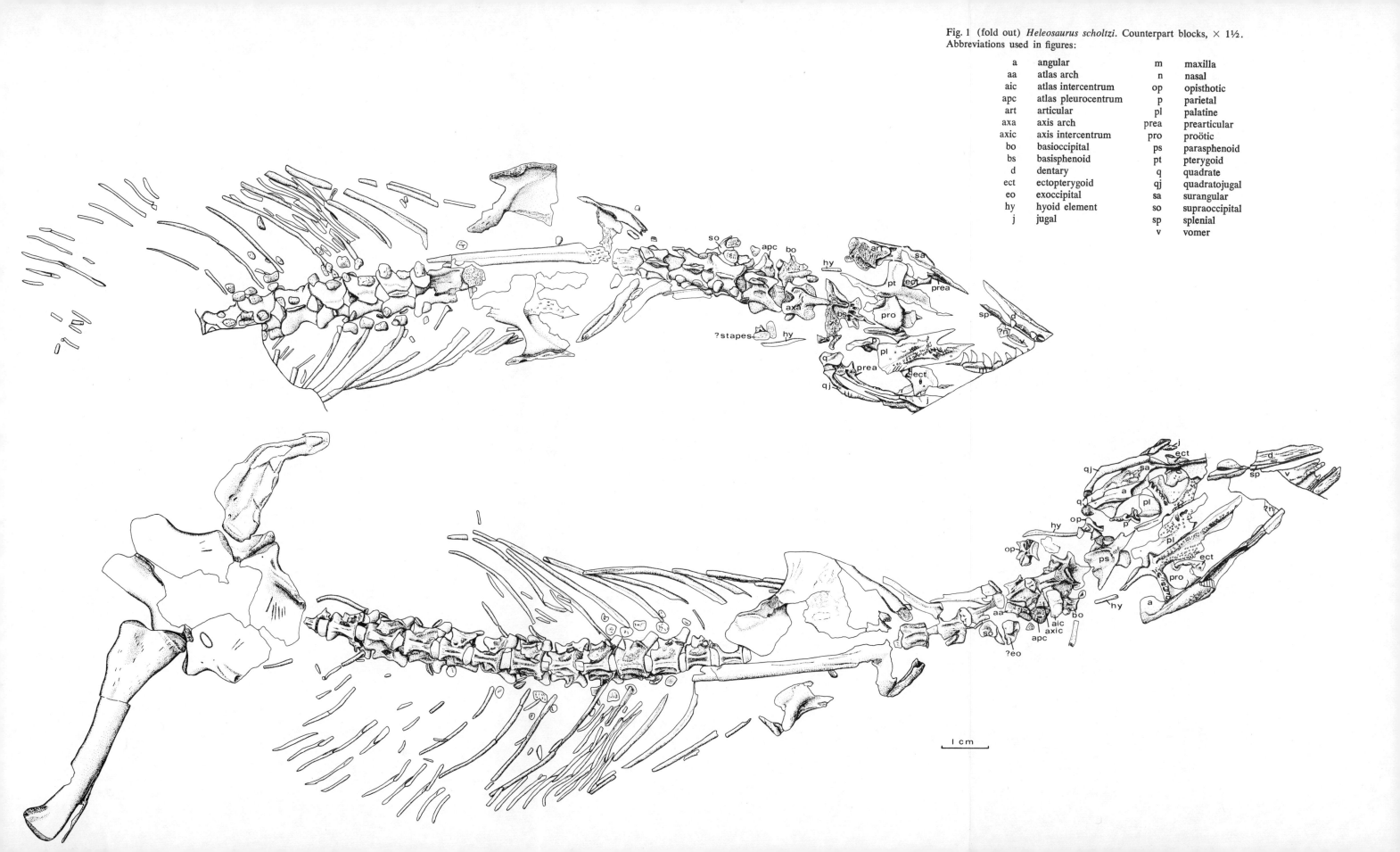

Fig. 1 (fold out) *Heleosaurus scholtzi*. Counterpart blocks, × 1½.
Abbreviations used in figures:

a	angular	m	maxilla
aa	atlas arch	n	nasal
aic	atlas intercentrum	op	opisthotic
apc	atlas pleurocentrum	p	parietal
art	articular	pl	palatine
axa	axis arch	prea	prearticular
axic	axis intercentrum	pro	proötic
bo	basioccipital	ps	parasphenoid
bs	basisphenoid	pt	pterygoid
d	dentary	q	quadrate
ect	ectopterygoid	qj	quadratojugal
eo	exoccipital	sa	surangular
hy	hyoid element	so	supraoccipital
j	jugal	sp	splenial
		v	vomer

1 cm

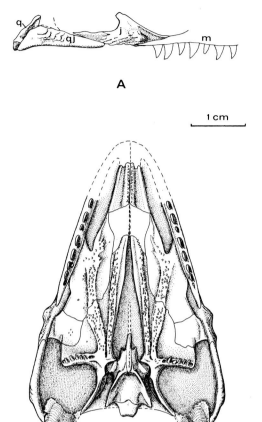

Fig. 2 *Heleosaurus scholtzi*, × 1½.

  A. Cheek region. Quadrate, quadratojugal, and jugal drawn as preserved with quadrato-
     jugal disarticulated slightly in an anteroventral direction from the quadrate and
     ventrally from the jugal. Maxilla is drawn as a mirror image outline from the
     medial surface. A groove in the dorsal surface of the quadratojugal indicates the
     margin of the squamosal.
  B. Restoration of palate.

for positive identification or meaningful description. A small, symmetrical plate
of bone adjacent to the fourth cervical is the supraoccipital. It is exposed in anterior view in the ventral block, and posteriorly in the dorsal. Its configuration
resembles that of *Euparkeria*, but it was not as solidly integrated into the skull
roof. The occipital surface as a whole probably resembled that of *Youngina*, but
the elements are not well enough known to attempt a restoration.

The bones of the otic capsule are well ossified, but disarticulated, and their
original orientation is difficult to establish. Both proötics can be seen in the area
of the transverse flange of the pterygoids. They are heavily ossified bones, showing
a groove for the passage of one or more branches of the fifth nerve.

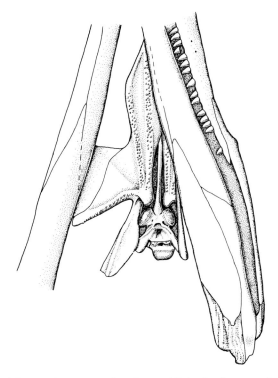

Fig. 3 Basicranial area of *Proterosuchus*. Bernard Price Institute specimen 3993, × ⅓.

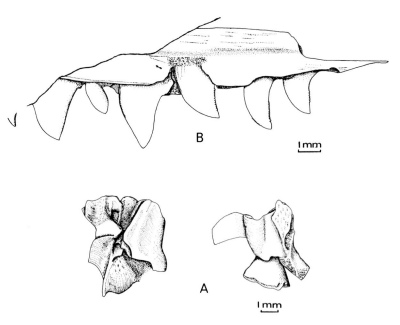

Fig. 4 *Heleosaurus scholtzi*.
A. Left and right opisthotics, × 4½.
B. Maxilla, × 6¼.

The right opisthotic is more or less in position, adjacent to the basisphenoid. The left lies adjacent to the intercentrum of the third cervical (Figs. 3, 4A). The disassociation and disarticulation of these elements make them almost impossible to compare with their counterparts in the very closely integrated braincase of *Euparkeria* or *Proterosuchus* (Cruickshank, 1971; 1972). No stapes has been positively identified, although an element in close association with the left opisthotic might be interpreted as the foot plate of a stapes with a closely placed stapedial foramen.

The right maxilla is nearly complete posteriorly, but approximately one-third of the anterior extent is lost (Fig. 4B). Part or all of seven teeth are exposed. Based on a comparison with *Euparkeria*, the complete maxillary tooth count was approximately 12. In detail, the teeth resemble those of that genus very closely. A portion of Ewer's (1965) description of that genus can be quoted directly in reference to *Heleosaurus*. "The teeth are laterally compressed, blade-like structures with sharp posterior and rather less distinct anterior keels. Both keels are finely serrated, but, while the serrations on the posterior keel are easily visible in most of the teeth, those on the anterior edge are more difficult to make out and can be clearly seen in only a few cases." Details of the tooth structure including the position of the serration in *Heleosaurus* can be noted in Figure 4B. Implantation of the teeth appears to be as fully thecodont as those of *Euparkeria*. In addition to the seven more or less fully erupted teeth, there are three replacement pits. The most posterior is just behind the last tooth, the second, with a tiny tooth in place, is adjacent to the second from the rear, and the third just behind the fourth. The third tooth from the front appears more conical than blade-like. It is shorter than the remainder and when fully erupted might have presented a somewhat more blade-like appearance.

Except for the specific nature of the marginal dentition, the portion of the skull that is preserved is remarkably similar to that of *Youngina* in size, proportions and details of bone configuration.

LOWER JAWS

Except for the area of the symphysis, most of the inner and outer surface of the jaws can be restored. The jaws are of the same general configuration as those of romeriids, except for the presence of a short retroarticular process. They also differ in having only a single splenial and apparently no coronoid, as is the case in *Euparkeria*. In contrast to that genus, there is no mandibular foramen. No more than approximately one-half of the mandibular margin of the left dentary is visible. Three fully erupted teeth are visible; the most posterior can only be seen in medial view, and so is not seen in the illustration of the dorsal surface. A further tooth at the anterior end of the exposed portion is in the process of erupting. These teeth generally resemble those of the upper jaw in being blade-like and sharply pointed, but no serrations are visible on either keel. The teeth are also slightly shorter and narrower. The same size relationship of upper and lower teeth applies to *Euparkeria*.

As in captorhinomorphs, there is a large articular, surrounded medioventrally by the surangular that supplied a large surface for the attachment of the pterygoideus musculature. The articulating surface is divided into two longitudinal depressions, angled several degrees medial to the long axis of the skull. There is a short but distinct retroarticular process, marked posterodorsally by numerous pits for muscle attachment. The angular has a rather limited lateral exposure, but extends anteriorly more than one-half the length of the jaw. The anterior portion of the surangular is raised into a slight coronoid prominence. This bone is marked laterally by a number of rounded processes. It is difficult to determine a function for these structures since their position is too far dorsal and anterior for them to have served for points of attachment of the pterygoideus musculature. There is no reason to think that the adductor musculature in this form extended onto the lateral surface of the jaw. The anterior end of the Meckelian fossa is slightly anterior to the end of the subtemporal fossa. Much of the inside of the jaw is formed by the single splenial that extends to the ventral margin to be visible for only a very short distance in lateral view.

Paired, rod-like hyoid elements are visible extending behind the quadrate rami of the pterygoids. Similar elements have been described in *Euparkeria*, *Proterosuchus* and *Prolacerta* (Camp, 1945).

VERTEBRAE AND RIBS

Seven vertebrae are visible anterior to the shoulder girdle, six of which may be considered cervicals on the basis of the configuration of the ribs (Fig. 5A). There is space for three vertebrae at the level of the shoulder girdle. This area cannot be prepared without damage to the girdle however. Thirteen more vertebrae are visible between the girdles (Fig. 5B). Initially two more presacral vertebrae were probably present anterior to the ilium but the posterior portion of the dorsal block is missing behind the 18th vertebrae. Therefore there may originally have been 25 presacrals. Judging by all the closely related forms, there were almost certainly two sacrals, although no evidence is provided by this specimen. Similarly, *Heleosaurus* probably had a long thin tail, but of this no trace remains.

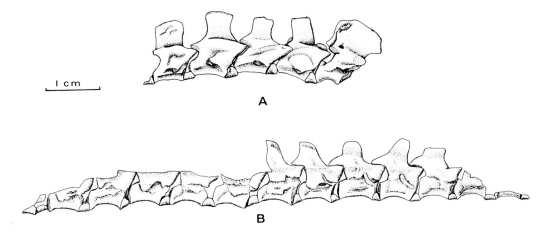

Fig. 5 *Heleosaurus scholtzi*. Vertebrae, × 1½.
A. Cervicals, numbers 2 – 6.
B. Trunk, numbers 11 – 23.

The atlas is completely disarticulated, but all the elements can be identified except for the proatlas. The intercentrum, pleurocentrum and arches, as well as the axis intercentrum lie beside the centrum of the axis and third cervical. The restoration can be seen in Figure 6. The exact lateral outline of the atlas centrum is difficult to determine as it is almost completely surrounded by other bone. It is obviously a large element, but most surfaces are of unfinished bone. suggesting that it did not extend ventrally or laterally to the level of the other central elements. This configuration presages the typical archosaurian condition in which this centrum is reduced to an odontoid, incorporated with the centrum of the axis. The intercentrum of the atlas is a widely crescentic bone, much larger than its counterparts in the remainder of the column. The median ventral surface forms a rounded keel. There are no obvious points for costal articulation at the posterior margin of the bone. Surprisingly, the two halves of the arch of the atlas have remained together, even though they have been separated from the remainder of the vertebra and lie three segments behind their normal position. They can be viewed ventrally, and to a limited extent laterally and medially. Their structure and relative position resemble those of *Euparkeria* and *Proterosuchus-Chasmatosaurus*.

In lateral view the intercentrum of the axis is a large bone, sharply angled ventrally and close to the shape of an equilateral triangle. In the large size of this element, *Heleosaurus* resembles the archosaurs and is in strong contrast to most captorhinomorphs in which this bone is incorporated into the centrum of the atlas. The arch and centrum of the atlas are solidly fused, with the spine, large and hatchet-shaped, overhanging much of the atlas. The posterior margin is deeply grooved for the attachment of interspinous ligaments. The base of the spine is indented posteriorly by deep, paired pits. The ventral margin of the centrum is bluntly keeled. In lateral view this margin is sharply concave ventrally. This centrum, like those in the remainder of the column, is notochordal. The transverse

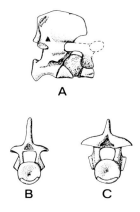

A

B   C

Fig. 6 *Heleosaurus scholtzi*. Restoration of vertebrae, × 2.
    A. Atlas-axis complex.
    B. 5th cervical in posterior view.
    C. 15th trunk vertebra in posterior view.

process is an ill-defined projection, extending ventrolaterally anterior to the middle of the centrum.

The neural spines of the remaining cervicals are rectangular, neither as long nor as wide relative to the size of the centrum as those of *Euparkeria* or *Proterosuchus*, but of generally similar configuration. The ill-defined transverse processes resemble closely those of *Euparkeria*. The intercentra are low-sided crescents. Their length at the midline is approximately one-half that of the centra. The fifth intercentrum shows well-developed areas for the articulation of the capitulum of the ribs on the posterior margin. Such are not evident on the more anterior elements. The sixth and seventh intercentra are not sufficiently well preserved to determine the presence of such articulatory surfaces.

In their overall configuration, the cervical vertebrae of *Heleosaurus* resemble those of primitive archosaurs, particularly *Euparkeria*, and are clearly distinct from those of late Pennsylvanian and early Permian romeriids.

Little can be seen of vertebrae 7 through 12. The 13th through 18th are well exposed (Fig. 5). They resemble those of *Youngina* (see Watson, 1957, p. 372), and also those of the genus *Captorhinus*, in having broad "swollen" neural arches, relatively short neural spines and primitive transverse processes. These vertebrae resemble also those at the rear of the cervical series in *Euparkeria*, but are quite unlike the distinctive trunk vertebrae that characterize archosaurs in general. The zygapophyses here, as in the rest of the column, are lateral in position and essentially flat. Although the lateral surfaces of all of the posterior vertebrae are somewhat distorted as a result of dorsoventral compression about the neural canal, it is apparent that the transverse processes are short projections angled anteroventrally, extending from the middle of the pedicel down towards the intercentrum. Intercentra 11 to 23 are all simple, low-sided crescents, none of which shows any specialized area for the articulation of the capitulum. Nothing is known of the more posterior vertebrae. The lengths of the centra remain essentially constant throughout the column.

The atlanteal ribs cannot be identified. Those of the axis are represented only by the distal end of the shaft of the left. It is narrow and flat, with the end fimbriated like the teeth of a comb. The head of the left third (Fig. 7A) is preserved, flattened into the centrum. Anteriorly, it shows a low ridge, between the tubercular and capitular heads, closely resembling that seen in *Euparkeria*. The ribs of the fifth vertebra have not been identified. Those of the fourth and sixth have long, flat blades, thickened dorsally, and clearly separate tubercular heads. The tubercular head of the sixth has a curious posteroventrally directed process. Presumably it was analogous with the extra processes seen in the cervical ribs of both *Euparkeria* and *Proterosuchus*, although it does not resemble them in detail.

The head of the seventh rib is not exposed, but a portion of the shaft can be seen extending from beneath the shoulder girdle. It resembles those more posteriorly in being cylindrical, rather than flattened as are the preceding cervicals. None of the heads of the trunk vertebrae can be clearly seen. The proximal end of the shaft is expanded, but it cannot be determined whether there were separate tubercular and capitular heads. The adequately preserved proximal ends of the shafts have rounded dorsal margins that are distinct (particularly posteriorly) from narrower ventral portions. The length of the trunk ribs increases to a maximum in

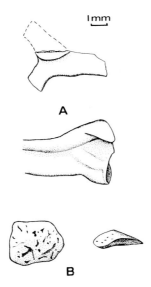

Fig. 7 *Heleosaurus scholtzi,* × 6.
    A. Anterior view of 3rd and posterior view of 6th rib.
    B. Dorsal and lateral views of dermal scales.

the area of the 14th, and then diminishes gradually toward the rear. None behind the 22nd is preserved. This rib is approximately one and one-half times the length of a centrum.

GIRDLES AND FEMUR

The shoulder girdle is preserved in more or less its original three-dimensional configuration, although ventrally it lies directly on the vertebral column. The general appearance resembles more that of captorhinomorphs than archosaurs other than *Proterosuchus.* The cleithrum was not observed, and may well not have been present. The clavicles are not completely exposed, but they are of the heavy build commonly seen in more primitive reptiles, and again as in *Proterosuchus.* The preserved portion of the stem reaches nearly to the top of scapula. A posterior flange extends from the lower part of the stem and continues onto the broad ventral plate. The interclavicle has the usual T-shape common to primitive reptiles. The anterior plate is poorly exposed, but apparently the anterior margin slopes sharply dorsally, as in early lepidosaurs and archosaurs. The stem is narrow and extends the length of five centra. The distal end is divided as in some romeriids and pelycosaurs.

The scapulocoracoid is ossified as a single unit. The blade is very short relative to the anterior-posterior extent of the coracoid area. A large subscapular fossa lies on the inner surface. The dorsal surface of the blade is thickened and rugose, suggesting continuation in cartilage. The posterior margin of the bone is thickened and well defined, but anteriorly the bone thins and its extent is difficult to ascertain. The scapulocoracoid resembles that of captorhinomorphs except for the anterior position of the coracoid foramen and the absence of the supraglenoid foramen. In these features it resembles both primitive archosaurs and millerettids. The glenoid is of the primitive screw-shape, retained also in *Proterosuchus*, necessitating a nearly horizontal orientation of the humerus. No feature of the shoulder girdle definitely presages the condition seen in more advanced archosaurs, but there is nothing to preclude such derivation either. Nothing of the forelimb is preserved.

The puboischiadic plate of the pelvic girdle is visible in ventral view, along with the base of the right ilium. This area generally resembles that of *Proterosuchus* (except for the further anterior extension of the pubis at the symphysis in *Heleosaurus*), but is really not far removed from the typical captorhinomorph pattern. Advanced features include the larger size of the obturator foramen, the prominence of the pubic tubercle, and the slight down-turning of the pubis. The pubis, ischium, and ilium apparently contribute equally to the acetabulum, unlike the condition in *Proterosuchus*, in which the pubis is excluded. The acetabulum is also more open ventrally than in that genus, and the ilium appears to extend laterally above the articulating surface. The posterior margin of the ischium is not thickened.

The entire left femur is preserved and appears in ventral view. The proximal end of the right is visible in the same orientation. The femur is far advanced from the captorhinomorph configuration, closely resembling that of crocodilians. The proximal end is flattened at approximately a 90° angle to the plane of the distal condyles. The head occupies only the anterior half of the proximal surface, and is angled slightly medially from the long axis of the shaft. Posteriorly there is a large area of roughened bone for the attachment of the M. puboischiofemoralis. The shaft is essentially straight. The area occupied by the intertrochanteric fossa in more primitive reptiles is nearly flat in *Heleosaurus*. There is neither an internal nor fourth trochanter. The ventral ridge system has vanished.

Despite the generally primitive configuration of the pelvic girdle, the anatomy of the femur indicates that the limb could most probably have been moved into a nearly vertical position, as Ewer (1965) has pointed out for the similar femur of *Euparkeria*. Nothing remains of the lower limb.

ARMOUR

Much of the ventral armour is preserved between the girdles. The individual ventral scales and their manner of articulation resemble those of romeriids and pelycosaurs, but also correspond well with those seen in early thecodonts. The most medial portion of the scales is not visible. There are at least four pairs of scales articulated in each row, and roughly 2.5 rows of scales per vertebra.

Dorsal dermal armour (Fig. 7B) is seen the length of the column, except where the vertebrae are covered by the shoulder girdle. There are at least 63 elements preserved for the 20 exposed vertebrae. This is clearly only a minimum number since most are seen in only a single plane. The plates vary from ovals as small as 1 mm in diameter, up to slightly irregular rectangles 3 mm in their longest dimension. Each is sculptured dorsally with an irregular pattern of pits. Most are rounded dorsally, although some are sharply keeled. None shows evidence of overlap. The ventral surface is flat to slightly concave. If the scales were contiguous, but without overlap, approximately 2.5 would occupy the length of a segment. The number preserved indicates that there must have been more than a single paired row along the vertebral column. Although these scutes are smaller and more numerous than those reported in any primitive thecodont, they follow the general pattern seen in that group, and can clearly be seen as antecedent to those of *Euparkeria*.

Discussion

BIOLOGY

The single known specimen of *Heleosaurus* is that of a small animal, 26 cm long from the posterior edge of the pelvic girdle to the tip of the snout. The tail, to judge from younginids currently being described, was probably at least this long again. The body is thus approximately half the size of *Euparkeria*, and far smaller than that of other primitive thecodonts. The degree of ossification of the girdles, the vertebrae, and the ends of the femur suggest that the animal was essentially mature when it died, although the elements of the braincase have become separated in a manner indicating that ossification was not as complete as that seen in *Euparkeria*. In the skeletal reconstruction (Fig. 8) the missing parts of the skull are restored according to the pattern of *Youngina*, without an antorbital fenestra.

In addition to the skeletal reconstruction, sketches have been made of the animal in several conceivable poses (Figs. 8, 9). The length ratios of the femur to humerus and proximal to distal limb bones are similar in other younginids and *Euparkeria* (Table 1) and those of the former group have been applied to *Heleosaurus*. There is little evidence, even indirect, of the probable proportions of the hands or feet. Neither is adequately known in younginids, but they appear to be large and primitive, as in captorhinomorphs, in marked contrast to the short hands and feet of primitive thecodonts.

The orientation of the glenoid in *Heleosaurus* indicates that the humerus must have been held in a primitive position, horizontal. This genus may have commonly assumed the sprawling posture of primitive reptiles. The configuration of the pelvic girdle and femur, however, is such that the rear limb could have been held vertically, or nearly so. If the rear limbs were held under the body, the forequarters would have had to be much lower because of the required pose of the humerus, unless a bipedal posture was assumed. This suggests that *Heleosaurus* probably ran in a bipedal manner.

Table 1. Limb measurements and proportions of younginids and *Euparkeria*, in mm. H: humerus; r: radius; U: ulna; F: femur; t: tibia; f: fibula. Measurements of *Youngina* courtesy of Chris Gow, based on Bernard Price Institute specimen 3859. *Galesphyrus* measurements based on the type, South African Museum specimen 2758. Measurements of *Euparkeria* from Ewer (1965) based on the type.

	H	r	u	$\frac{r}{H}$ %	C F	t	f	$\frac{t}{F}$ %	$\frac{H}{F}$ %
Youngina capensis	23	18	16	75	33.5	—	—	—	69
Galesphyrus capensis	21	17	16.5	81	27	25	26	92	78
Euparkeria capensis	37.8	31.8	33.7	84	55.8	47.8	—	86	66

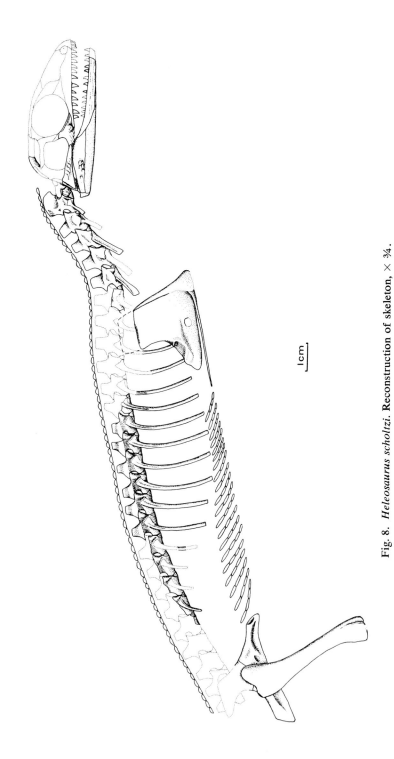

1cm

Fig. 8. *Heleosaurus scholtzi.* Reconstruction of skeleton, × ¾.

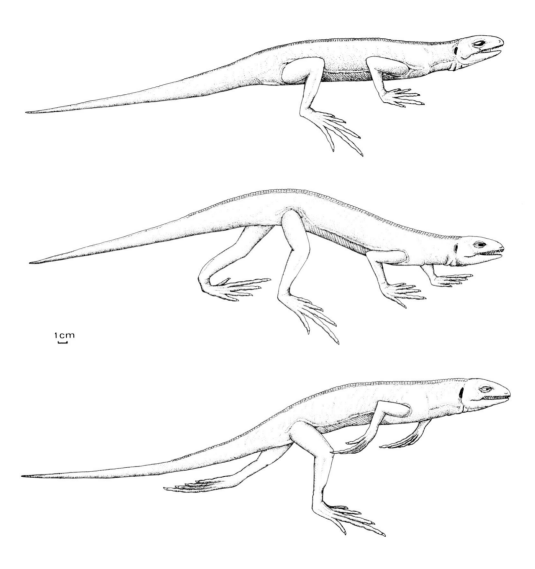

Fig. 9 *Heleosaurus scholtzi*. Restoration in three poses. Tail and lower limbs restored on the pattern of *Youngina* and *Galesphyrus*, × ¼.

In the absence of the lower limb elements and the tail, it is not possible to pre-clude the possibility that *Heleosaurus* showed some specializations toward aquatic locomotion. On the other hand the high degree of ossification and the overall similarity (particularly of the dentition) to *Euparkeria* strongly suggest that *Heleosaurus* was an agile terrestrial carnivore.

Comparisons of *Heleosaurus* with Primitive Thecodonts

Throughout this description, comparisons have been made between *Heleosaurus* and both *Euparkeria* and *Proterosuchus-Chasmatosaurus*, the best known primi-tive thecodonts. In the retention of primitive characteristics, most notably seen in the girdles, this genus is similar to *Proterosuchus*. In terms of general habitus, however, the similarity to *Euparkeria* is significant. Size, dentition, body propor-tions and possibly posture indicate that *Heleosaurus* and *Euparkeria* may have enjoyed a very similar way of life.

In terms of the relative age of their fossil remains (Fig. 10) and a simple cata-logue of skeletal characteristics, one might arrange these forms in the sequence *Heleosaurus-Proterosuchus-Euparkeria*. The obviously inadequate record of early thecodonts, because of the rarity of appropriate ancestors for the more advanced groups, indicates that any specific phylogenetic arrangement of the few well-known forms is almost certain to be incorrect in detail. The evidence provided by *Heleo-saurus* suggests rather an early dichotomy—*Euparkeria* representing a further sophistication of the habitus of an essentially primitive agile, terrestrial carnivore, while *Proterosuchus* represents a lineage that became specialized at an early stage toward large size, with accompanying aquatic specialization and pisciverous habits.

The early divergence of *Proterosuchus-Chasmatosaurus* and *Erythrosuchus* from euparkeriids has been suggested by other authors, and Watson's phylogeny (1957, fig. 23) indicates a dichotomy of two thecodont stocks at the eosuchian level.

Origin of Archosaurs

The presence of a generally thecodont pattern of armour in *Heleosaurus*, the simi-larity of the dentition of this genus with that of *Euparkeria*, and the resemblance of the cervical vertebrae and girdles to *Proterosuchus* give this genus a far greater similarity to primitive archosaurs than that shown in any other taxon.

Unfortunately our knowledge of the characteristic eosuchian, *Youngina*, is much less complete than is that of early archosaurs. Nevertheless, the portions of the anatomy known in common indicate that *Heleosaurus* is an eosuchian and may be retained in the family Younginidae, although it occurs in a lower zone within the Beaufort than that from which all the specimens of *Youngina* have been recovered (Kitching, 1971). Thus it may be said with considerable confidence that the archosaurs are closely related to eosuchians, and it may be inferred that the two groups have a fairly simple ancestor-descendant relationship.

The suggestions by Romer (1967) and Reig (1970) that archosaurs evolved from groups other than eosuchians are both based on, among other things, the premise that *Proterosuchus-Chasmatosaurus* is an appropriate representative of the primitive thecodont stock. Romer has emphasized the posterior position of the jaw suspension and the apparent absence of the otic notch as precluding derivation of the group from younginids, which have a definite otic notch and the jaw articulation at the level of the occipital condyle. Cruickshank (1972) has clearly demonstrated that *Proterosuchus* does have an otic notch, comparable to that of *Euparkeria* but also readily derivable from that seen in *Youngina*. Further, the mobility of the quadrate in this genus is such that the jaw articulation may be moved into a position nearly opposite the occipital condyle. Comparison of the palates of a

			YOUNGINIDS	ARCHOSAURS
TRIASSIC	Stormberg series			
	Cynognathus zone			EUPARKERIA ERYTHROSUCHUS
	Lystrosaurus zone			PROTEROSUCHUS
PERMIAN	Daptocephalus zone		YOUNGINA	ARCHOSAURUS
	Cistecephalus zone		HELEOSAURUS GALESPHYRUS	
	Tapinocephalus zone			
	Ecca series			
	Dwyka series			
PENNSYLVANIAN	Stephanian		PETROLACOSAURUS	

Fig. 10 Geological occurrence of genera mentioned in text. The relative thickness of the units is arbitrary.

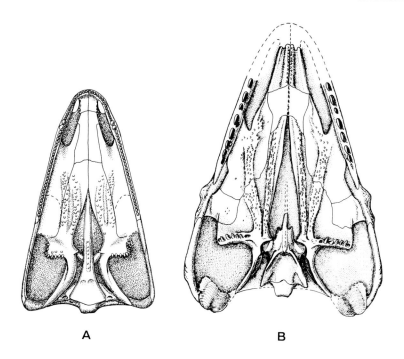

A   B

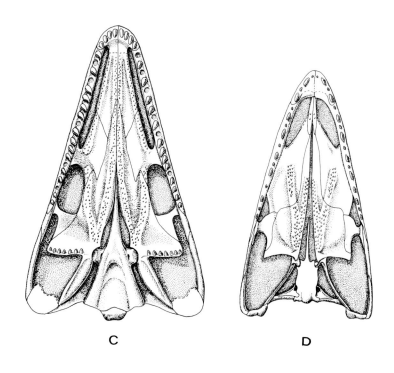

C   D

romeriid, *Youngina, Heleosaurus*, and *Euparkeria* (Fig. 11) demonstrates the very conservative nature of all of the important features. The presence of a "short" jaw in most advanced archosaurs demonstrates that *Proterosuchus* is an exception to this pattern, but the nature of this genus certainly does not preclude the evolution of archosaurs from "short-jawed" eosuchians.

The apparently aquatic habits of *Proterosuchus-Chasmatosaurus* were emphasized by Reig (1970) in his discussion of the ancestry of archosaurs. Charig (pers. comm.) has also considered such a habitat as being important in relationship to the origin of dinosaurs. There seems no reason to dispute the aquatic nature of this genus nor of *Erythrosuchus*, yet there is no clear evidence that these forms gave rise to any more advanced thecodonts. On the basis of their anatomy, they do not appear to represent the basal stock, but rather (in a fashion similar to the ichthyostegids) exemplify an early specialization in quite a different direction to that assumed by the eosuchian ancestors or the central stock of the pseudo-suchians, represented by *Euparkeria*.

Origin of Eosuchia

A further topic, largely beyond the bounds of this paper, should be mentioned. If the eosuchians are ancestral to archosaurs, from what group did the eosuchians themselves evolve? A study now nearing completion by T.H. Eaton and R. Reisz substantiates Peabody's (1952) suggestion that *Petrolacosaurus kansensis* from the Upper Pennsylvanian of Kansas has a diapsid skull and is, in general, a plausible ancestor for the late Permian diapsids. This assumes a long hiatus in the fossil record between the first appearance of diapsids and their differentiation in the late Permian. The absence of fossils in the intervening period may be partially explained by assuming that this lineage was primarily terrestrial in habits, so its members were unlikely to be fossilized in the predominantly deltaic deposits of the Lower Permian of Europe and North America.

Fig. 11 Palates.
    A. The romeriid *Paleothyris*, × 2.
    B. *Heleosaurus*, × 1½.
    C. *Youngina*, × 1½.
    D. *Euparkeria*, × ¾. Base of braincase in *Euparkeria* is not well exposed in any specimens. Otic capsule of *Youngina* and *Heleosaurus* not restored. *Youngina* is based on American Museum specimen 5567 and Field Museum specimen 1528. *Euparkeria* is redrawn from Ewer (1965), with details of vomers incorporated from Gow (1970).

Acknowledgments

Permission to study the specimen that forms the basis for this paper was given by Dr. Michael Cluver of the South African Museum. I also wish to express my appreciation to the director, Dr. T.H. Barry for providing facilities during my stay at that museum. Technical assistance there was given by Mr. Neville Eden. Dr. A.R.I. Cruickshank and Mr. C. E. Gow at the Bernard Price Institute, University of the Witwatersrand were most helpful in providing information on their work in progress on eosuchians and archosaurs and in discussing the subject of this article. The help of Dr. J.W. Kitching in establishing the horizon in which *Heleosaurus* and *Galesphyrus* were found is greatly appreciated. This work was supported by grants from the National Research Council of Canada and the Faculty of Graduate Studies and Research of McGill University.

Literature Cited

BROOM, R.

1907   On some new fossil reptiles from the Karroo beds of Victoria West, South Africa. Trans. S. Afr. Phil. Soc., 18: 31–42.

CAMP, C.L.

1945   *Prolacerta* and the protorosaurian reptiles. Amer. J. Sci., 243: 17–32, 84–101.

CRUICKSHANK, A.R.I.

1971   Early thecodont braincases. *In* Second Gondwana Symposium, proceedings and papers. Pretoria, Council of Scientific and Industrial Research, pp. 683–685.

1972   The proterosuchian thecodonts. *In* Joysey, K.A. and T.S. Kemp, eds., Studies in Vertebrate Evolution. Edinburgh, Oliver and Boyd, pp. 89–119.

EWER, R.F.

1965   The anatomy of the thecodont reptile *Euparkeria capensis* Broom. Phil. Trans. Roy. Soc. London (B), 248: 379–435.

GOW, C.E.

1970   The anterior of the palate in *Euparkeria*. Palaeont. Afr., 13: 61–62.

1972   The osteology and relationships of the Millerettidae (Reptilia: Cotylosauria). J. Zool., Lond., 167: 219–264.

HUGHES, B.

1963   The earliest archosaurian reptiles. South Afr. J. Sci., 59: 221–241.

KITCHING, J.W.

1971   A short review of the Beaufort zoning in South Africa. *In* Second Gondwana Symposium, proceedings and papers. Pretoria, Council of Scientific and Industrial Research, pp. 309–312, plus map.

PEABODY, F.E.

1952   *Petrolacosaurus kansensis* Lane, a Pennsylvanian reptile from Kansas. Univ.
        Kansas Palaeont. Contrib., Vertebrata, Article 10: 1–41.

REIG, O.A.

1967   Archosaurian reptiles: a new hypothesis on their origins. Science, 157: 565–568.

1970   The Proterosuchia and the early evolution of the archosaurs; an essay about the
        origin of a major taxon. Bull. Mus. Comp. Zool. Harv., 139(5): 229–292.

ROMER, A.S.

1956   Osteology of the reptiles. Chicago, Univ. Chicago Press, 772 pp.

1966   Vertebrate Paleontology. 3rd ed., Chicago, Univ. Chicago Press, 468 pp.

1967   Early reptilian evolution reviewed. Evolution, 21(4): 821–833.

1971   Unorthodoxies in reptilian phylogeny. Evolution, 25(1): 193–112.

TATARINOV, L.P.

1961   [Material on the USSR pseudosuchians.] Palaeont. Zh. 1: 117–132 (in Russian).

WATSON, D.M.S.

1957   On *Millerosaurus* and the early history of the sauropsid reptiles. Phil. Trans. Roy.
        Soc. London (B), 240(673): 325–400.

Cenozoic Herpetofaunas
of Saskatchewan

J. Alan Holman
Curator of Vertebrate Paleontology, The Museum,
Michigan State University, East Lansing

Abstract

Before the 1960s very little was recorded about the Tertiary amphibians and reptiles of Canada. But during the 1960s vertebrate collecting in the Cypress Hills and Wood Mountain Formations of Saskatchewan produced some small assemblages of fossil amphibians and reptiles that have evolutionary and palaeoclimatic significance. An essentially modern amphibian fauna, but a somewhat archaic reptile fauna, occurred in Saskatchewan in the early Oligocene. A distinct modernization of the reptile fauna took place between early Oligocene and late Miocene times. Between late Miocene and Recent times many members of the earlier faunas retreated to the south, reflecting changes from a Tertiary sub-tropical climate with a mesophytic deciduous forest vegetation to a Recent north temperate climate with a xerophytic grassland vegetation. The earliest fossil records for eight Recent United States genera are from the Tertiary of Saskatchewan.

Introduction

E.D. Cope recorded the first Tertiary reptiles from Saskatchewan in 1891 based on material collected in 1884. Later, Lambe (1906 and 1908) recorded a few other reptiles, also from the Cypress Hills. L.S. Russell (1934b) reported on two turtles from the Ravenscrag Formation (Palaeocene) of Saskatchewan (*Trionyx* cf. *T. subquadratus* Lambe, 1914, and *Clemmys backmani* Russell, 1934b). In the 1960s, collecting in Saskatchewan, especially by the Royal Ontario Museum, the National Museum of Canada, and the Saskatchewan Museum of Natural History, brought small amphibian and reptile fossils to light. Estes (1970) and Holman (1963, 1968, 1970, 1971, and 1972) have discussed these remains.

The present paper summarizes our present information concerning Tertiary herpetofaunas of Saskatchewan, and compares the fossil herpetofaunas to the Recent herpetofauna of the province. The study of Canadian Tertiary amphibians and reptiles is yet in its infancy, and the most important works on Canadian Cenozoic amphibians and reptiles are yet to come. As Estes (1970, p. 139) states, "The vertebrate fossil record, with paleogeographical and paleobotanical data, indicates that Canada was the site of development of many present-day temperate forms. Since few continental Tertiary localities occur in the eastern United States, an area where a temperate lower vertebrate fauna is now widely distributed, it is clear that the history of this fauna must be sought in Cretaceous and Cenozoic deposits to the north."

Following are checklists of the three known Cenozoic herpetofaunas of Saskatchewan; an early Oligocene one, a late Miocene one, and the Recent herpetofauna.

Checklists of Cenozoic Herpetofaunas of Saskatchewan

Early Oligocene

CYPRESS HILLS FORMATION NEAR EASTEND, SASKATCHEWAN

Class Amphibia
  Family Ambystomatidae
    Ambystoma tiheni Holman, 1968 Extinct Mole Salamander
  Family Rhinophrynidae
    Rhinophrynus canadensis Holman, 1963 Extinct Cone-nosed Toad
  Family Pelobatidae
    Scaphiopus (Scaphiopus) skinneri Estes, 1970 Extinct Spadefoot Toad
  Family Hylidae
    Hyla swanstoni Holman, 1968 Extinct Treefrog
Class Reptilia
  Family Carettochelyidae
    Anosteira ornata ? Leidy, 1871 Extinct Anosteirine Turtle
  Family Trionychidae
    Trionyx leucopotamicus Cope, 1891 Extinct Softshell Turtle
  Family Testudinidae
    Stylemys nebrascensis Leidy, 1851 Extinct Land Tortoise
    Geochelone exornata (Lambe, 1906) Extinct Land Tortoise
  Family Emydidae
    Emydidae indeterminate Pond Turtle
  Family Crocodilidae
    Crocodylus prenasalis ? Loomis, 1904 Extinct Crocodile
  Family Amphisbaenidae
    Leostophis near *L. anceps* (Marsh, 1871) Large Extinct Worm Lizard
    Amphisbaenidae indeterminate Small Extinct Worm Lizard
  Family Iguanidae
    Cypressaurus hypsodontus Holman, 1972 Extinct Sceloporine Lizard
    Crotaphytus oligocenicus Holman, 1972 Extinct Collared Lizard
  Family Anguidae
    Glyptosaurus sp. Extinct Glyptosaurine Lizard
    Peltosaurus granulosus Cope, 1873 Extinct Glyptosaurine Lizard
  Family Xantusidae
    Palaeoxantusia borealis Holman, 1972 Extinct Night Lizard
  Family Boidae
    Ogmophis compactus Lambe, 1908 Extinct Medium-sized Boa
    Calamagras weigeli Holman, 1972 Extinct Small Boa

Late Miocene

WOOD MOUNTAIN FORMATION NEAR ROCKGLEN, SASKATCHEWAN

Class Amphibia
  Family Pelobatidae
     Scaphiopus (*Spea*) cf. *alexanderi* (Zweifel, 1956) Extinct Spadefoot Toad
  Family Bufonidae
     Bufo valentinensis Estes and Tihen, 1964 Extinct True Toad
Class Reptilia
  Family Testudinidae
     Geochelone sp. Extinct Giant Land Tortoise
  Family Emydidae
     Emydidae indeterminate Pond Turtle
  Family Crocodilidae
     Crocodilian indeterminate Small Crocodilian
  Family Iguanidae
     Sceloporus near *S. magister* Hallowell, 1854 Spiny Lizard
  Family Anguidae
     Ophisaurus canadensis Holman, 1970 Extinct Glass Lizard
  Family Boidae
     Charina prebottae Brattstrom, 1958 Extinct Rubber Boa
  Family Colubridae
     Paracoluber storeri Holman, 1970 Extinct Racerlike Snake
     Elaphe nebraskensis Holman, 1964 Extinct Ratsnake
     Lampropeltis similis Holman, 1964 Extinct Small Kingsnake
     Colubrinae indeterminate ? "Archaic Colubrinid Snake"

Recent Fauna

REVISED FROM LOGIER AND TONER, 1961

Class Amphibia
  Family Ambystomatidae
     Ambystoma tigrinum diaboli Dunn, 1940 Gray Tiger Salamander
     Ambystoma tigrinum melanostictum (Baird, 1860) Blotched Tiger Salamander
  Family Pelobatidae
     Scaphiopus (*Spea*) *bombifrons* Cope, 1863 Plains Spadefoot
  Family Bufonidae
     Bufo cognatus Say, 1823 Great Plains Toad
     Bufo hemiophrys Cope, 1886 Dakota Toad
  Family Hylidae
     Pseudacris triseriata maculata (Agassiz, 1850) Boreal Chorus Frog
  Family Ranidae
     Rana sylvatica Le Conte, 1825 Wood Frog
     Rana pipiens Schreber, 1782 Leopard Frog

Class Reptilia
  Family Chelydridae
     Chelydra serpentina serpentina (Linnaeus, 1758) Common Snapping Turtle
  Family Emydidae
     Chrysemys picta belli (Gray, 1831) Western Painted Turtle
  Family Iguanidae
     Phrynosoma douglasii brevirostre (Girard, 1858) Eastern Short-horned Lizard

Family Colubridae
Thamnophis elegans vagrans (Baird and Girard, 1853) Wandering Garter Snake
Thamnophis radix haydeni (Kennicott, 1860) Western Plains Garter Snake
Thamnophis sirtalis parietalis (Say, 1823) Red-sided Garter Snake
Heterodon nasicus nasicus Baird and Girard, 1852 Plains Hognose Snake
Opheodrys vernalis blanchardi Grobman, 1941 Western Smooth Greensnake
Pituophis melanoleucus sayi (Schlegel, 1837) Bullsnake
Crotalus viridis viridis (Rafinesque, 1818) Prairie Rattlesnake

Early Oligocene Herpetofauna

The early Oligocene is represented by a single herpetofauna in Canada (Checklist and Figs. 1 and 2) from the Cypress Hills Formation near Eastend, Saskatchewan. Fossil mammals from the Cypress Hills Formation indicate that it is approximately contemporaneous with the Chadron Formation of South Dakota (Russell, 1948). Recent herpetological collections have come from the Calf Creek local fauna along the north branch of Calf Creek about 10 miles northwest of Eastend. Other herpetological papers on this site are Holman (1963, 1968, and 1972). Other modern papers on the Cypress Hills Oligocene are those on the birds (Weigel, 1963) and on geology and mammals (Russell 1934a and b, 1936, 1938, 1940, and 1948). Dr. L.S. Russell has recently studied additional mammalian material from the Cypress Hills.

Evolutionary Aspects

The Cypress Hills Lower Oligocene amphibians and reptiles are modern at the family level, as all 13 families identified are extant. In contrast, over 50 per cent of the mammalian families from the deposit are extinct (Russell, 1948). The amphibian fauna is more modern than the reptilian fauna in that all four amphibian genera identified are extant, whereas only four of 13 reptilian genera are extant. Several modern genera now typical of the United States and Mexico have their earliest fossil occurrences in the Cypress Hills fauna. These are: *Ambystoma*, *Rhinophrynus*, *Scaphiopus* (subgenus *Scaphiopus*), *Hyla*, and *Crotaphytus*. Some of the Cypress Hills forms are tentatively assigned to Recent species groups and to lines leading to living species. These include *Ambystoma tiheni* (to Recent *A. opacum*), *Rhinophrynus canadensis* (to Recent *R. dorsalis*), *Scaphiopus skinneri* (to Recent *S. holbrooki*), and *Crotaphytus oligocenicus* (to Recent *C. collaris* and *C. wislizenii*).

A striking feature of the Cypress Hills herpetofauna is the absence of frogs of the genera *Bufo* (Bufonidae) and *Rana* (Ranidae). *Bufo* is first known from the Oligocene of South America and *Rana* is first known from the Eocene of Europe. *Bufo* and *Rana* are important elements of the fossil frog faunas from early Miocene through Pleistocene times in North America, and these genera each have large species complexes in North America today.

Another striking feature is that lizards are more abundant than snakes in the Cypress Hills fauna. At least four families of lizards representing at least seven species (probably more) are present, whereas only two species of one family

(Boidae) represent the snakes. Snakes outnumber the lizards in the Upper Miocene herpetofauna, and overwhelmingly outnumber the lizards in the Recent herpetofauna of Saskatchewan (see Checklists).

Palaeoecology

The Cypress Hills Formation lies in the ecotonal zone between the Temperate Unit of the Arcto-Tertiary flora to the North and the Neotropical-Tertiary flora to the south, based on palaeobotanical evidence (see Chaney, 1947; Axelrod, 1958; and map in Kendeigh, 1961, p. 283). The climate of the temperate unit of the Arcto-Tertiary flora is thought to have been humid with extensive summer rainfall, and with moderate temperatures that did not fall below freezing. The Neotropical flora to the south is assumed to have lived in an even more tropical climate. Many elements of the Cypress Hills herpetofauna indicate a subtropical climate. These include the Cone-nosed Toad, Anosteirine Turtle, Land Tortoises, Crocodile, Worm Lizards, and Boas.

Two amphibians (*Ambystoma tiheni* and *Scaphiopus skinneri*) may be indicative of deciduous forest conditions as at least a part of the vegetational complex of Saskatchewan during the early Oligocene, for *Ambystoma opacum* and *Scaphiopus holbrooki*, the most closely related living members of the fossil species, are

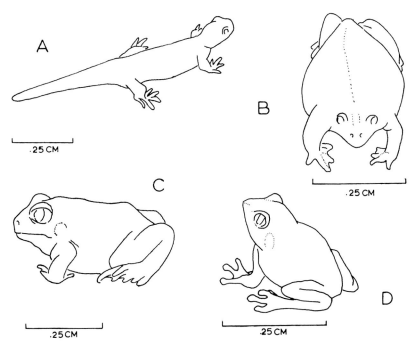

Fig. 1 Restorations of amphibians of early Oligocene Cypress Hills fauna, Saskatchewan. Drawings by Donna R. Holman.
  A. *Ambystoma tiheni.*
  B. *Rhinophrynus canadensis.*
  C. *Scaphiopus skinneri.*
  D. *Hyla swanstoni.*

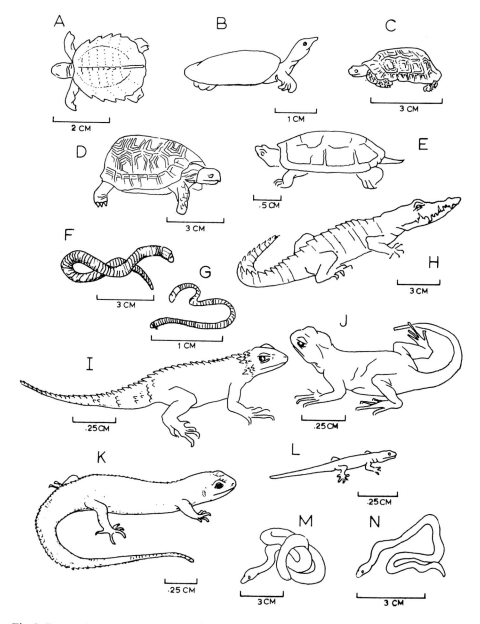

Fig. 2 Restorations of reptiles of early Oligocene Cypress Hills fauna, Saskatchewan. Drawings by Donna R. Holman.

A. *Anosteira ?ornata.*
B. *Trionyx leucopotamicus.*
C. *Stylemys nebrascensis.*
D. *Geochelone exornata.*
E. Emydidae indeterminate.
F. *Lestophis* near *anceps.*
G. Amphisbaenidae indeterminate.

H. *Crocodylus ?prenasalis.*
I. *Cypressaurus hypsodontus.*
J. *Crotaphytus oligocenicus.*
K. *Peltosaurus granulosus.*
L. *Palaeoxantusia borealis.*
M. *Ogmophis compactus.*
N. *Calamagras weigeli.*

almost entirely restricted to the eastern deciduous habitat in the United States today. But some of the Oligocene forms (*Crotaphytus oligocenicus*, *Palaeoxantusia borealis*, and *Calamagras weigeli*) evidently gave rise to forms that ultimately became adapted to arid conditions in the southwestern United States (*Crotaphytus collaris* and *wislizenii*; *Xantusia vigilis* and *arizonae*; and *Lichanura roseofusca*).

Stratigraphic Comparisons
The only large North American herpetofauna at all temporally near the early Oligocene Cypress Hills herpetofauna is the Tabernacle Butte herpetofauna of the Middle Eocene of the Bridger Formation of Wyoming (Hecht, 1959). As might be expected, there are several differences (some of which might be due to accidents of preservation) between the two faunas, as follows: (1) lack of salamanders, hylid frogs, turtles, and iguanid lizards in the Tabernacle Butte fauna (these groups are all present in the Cypress Hills fauna); and (2) the lack of diplasiocoel frogs, the rarity of anguid lizards, the lack of varanid, agamid, and teiid lizards, and the absence of aniliid snakes in the Cypress Hills fauna (these groups are all present in the Tabernacle Butte fauna).

Late Miocene Herpetofauna

The late Miocene is represented by a single herpetofauna in Canada (Checklist and Fig. 3) from the Wood Mountain Formation near Rockglen, Saskatchewan. Fossil mammals from this locality are comparable to those of Barstovian (Upper Miocene) localities elsewhere (John Storer, pers. comm.). Dr. John Storer completed a doctoral dissertation at the University of Toronto on small mammals from this deposit. This material was collected by Dr. Storer and Dr. L.S. Russell in 1967 and 1968. In 1970 Dr. L.S. Russell and party made extensive collections at the Kleinfelder Farm locality near Rockglen. Tons of concentrate were collected to be sorted through for fossils in future years. Holman (1970) reported on herpetological material from the 1967 and 1968 collections and in 1971 Holman reported on giant tortoise material collected in 1955 and in 1970. Additional studies on microfaunal remains from the concentrate collected by Dr. L.S. Russell should enlarge the list of all vertebrate classes from the site.

Evolutionary Aspects
By early Miocene times the modern nature of the herpetofauna is striking, as, with the exception of two colubrid snakes, all of the identifiable forms may be assigned to living genera, and many appear to be ancestral to living species groups. Both amphibian genera identified are modern and six of eight reptilian genera are living today. All forms identified to the specific level are extinct.

Some Wood Mountain forms that may at least tentatively be assigned to Recent species groups and to lines leading to living species include *Scaphiopus* (*Spea*) cf. *alexanderi* (to Recent *S. bombifrons*, *hammondi*, *intermontanus* group, subgenus *Spea*); *Bufo valentinensis* (to Recent *B. americanus* group); *Sceloporus* near *S. magister* (questionably to Recent *S. magister*); *Ophisaurus canadensis* (to

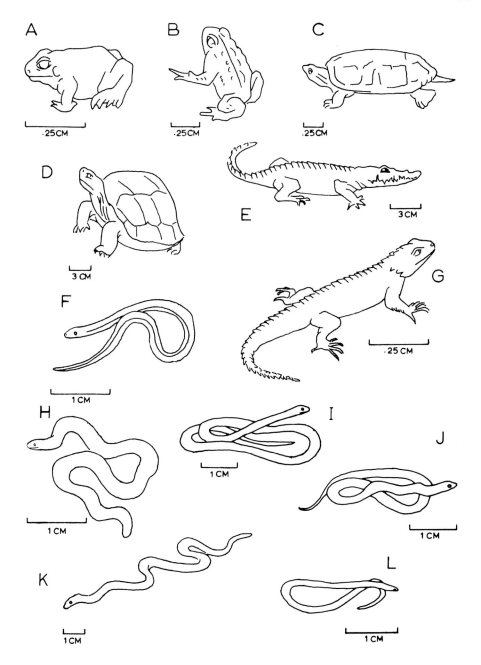

Fig. 3 Restorations of herpetofauna of late Miocene Wood Mountain fauna, Saskatchewan.
Drawings by Donna R. Holman.

A. *Scaphiopus* cf. *alexanderi*.
B. *Bufo valentinensis*.
C. Emydidae indeterminate.
D. *Geochelone* sp.
E. Crocodilian indeterminate.
F. *Ophisaurus canadensis*.

G. *Sceloporus* near *magister*.
H. *Charina prebottae*.
I. *Paracoluber storeri*.
J. Colubrinae indeterminate.
K. *Elaphe nebraskensis*.
L. *Lampropeltis similis*.

Recent *O. ventralis*); *Charina prebottae* (to Recent *C. bottae*); *Elaphe nebraskensis* (to Recent *E. vulpina*); and *Lampropeltis similis* (to Recent *L. triangulum*). This is the earliest fossil record for the genus *Sceloporus* and the earliest American fossil record for the genus *Ophisaurus*.

Palaeoecology

It is generally believed that with the elevation of the western mountains in the North American Miocene and Pliocene there was a deterioration of the climate that caused a withdrawal of the Neotropical and Tertiary flora from the north to its present location, and of the Temperate Unit of the Arcto-Tertiary flora to the southeastern United States. The Wood Mountain herpetofauna is of considerable interest in reflecting local climatic conditions during late Miocene times, especially since this fauna has species closely related to modern ones whose ecological tolerances are known.

From a distributional standpoint (1) many of the herpetological forms have living relatives that occur well south of the fossil locality, and (2) several of these southern forms have living relatives distributed to southwest or to the southeast of the fossil locality today. Group 1 forms are *Geochelone* sp., a crocodilian, *Sceloporus* sp. near *S. magister*, *Ophisaurus canadensis*, *Elaphe nebraskensis*, and *Lampropeltis similis*. Group 2 forms are *Sceloporus* sp. near *S. magister* (southwest), *Ophisaurus canadensis* (southeast), *Charina prebottae* (southwest), a crocodilian (southeast), and *Elaphe nebraskensis* (southeast). These distributional patterns were interpreted by Holman (1970) as an indication that in late Miocene times: "(1) the climate near the Rockglen area of Saskatchewan was much warmer than at present, especially in winter when there must have been only a few days of frost, and (2) ... perhaps east and west rainfall zonation in North America was not as pronounced as at present."

Then Holman (1971) reported on giant tortoise material, previously not recorded from the Wood Mountain Formation. This material was collected by W.L. Langston, Jr., and L.S. Russell in 1955 and by L.S. Russell in 1970. It was pointed out that these fossils suggest a subtropical climate in southern Saskatchewan during late Miocene times. This thesis is based on the following reasoning, first presented by Hibbard (1960). Hibbard stated that large tortoises of the genus *Geochelone* are restricted to tropical areas today, and he pointed out that shortly after colonies of living Galapagos Island *Geochelone* were brought to localities in southeastern and southwestern United States (San Diego, California; Superior, Arizona; Houston, and San Antonio, Texas; New Orleans, Louisiana; Brighton, Opa-Locha, Biscayne Island, and Lignum Vitae Island, Florida) it was found that the animals needed protection in winter at all of these localities, for the tortoises could not survive cold nights when the temperature was near or below 32°F (= 0°C). Hibbard states, "The large Pliocene and Pleistocene land tortoises could not stand freezing, but they might have existed in an area where very few light frosts occurred at night but with the temperature during the day warmed to 60 degrees or more Fahrenheit."

In the area where these large animals occurred there were no caves nor rock shelters where they could have retreated in cold weather, and the tortoises had stumpy, elephant-like feet that were unsuitable for burrowing. Moreover, to postulate that the giant tortoises could have migrated to southern lands, or that these cold-blooded animals that could not hibernate could have become "physiologically adjusted" to freezing temperatures seem most unlikely assumptions. Thus it seems reasonable to suggest a subtropical climate with winter temperatures not lower than 32°F ($= 0°C$) for southern Saskatchewan during late Miocene times. I cannot suggest exactly what type of flora was present in southern Saskatchewan during the late Miocene, but it certainly seems safe to suggest that it was more mesophytic and contained more southern species than the flora of today.

Stratigraphic Comparisons

The only large North American fossil herpetofauna that approaches the Wood Mountain Formation in age is that of the Norden Bridge Quarry fauna of Brown County, Nebraska. This deposit is thought to be essentially transitional in age between the Miocene and the Pliocene (Estes and Tihen, 1964). The Norden Bridge Quarry fauna has more herpetological species than the Wood Mountain fauna, and the Norden Bridge fauna has many aquatic and semiaquatic species; including ranid and hylid frogs, soft-shelled turtles, and natricine snakes that are not present in the Wood Mountain fauna. But three species from the Wood Mountain fauna were previously known only from the Norden Bridge fauna. These are *Bufo valentinensis*, *Elaphe nebraskensis*, and *Lampropeltis similis*. Elements from the Wood Mountain fauna that are conspicuously absent from the Norden Bridge Quarry fauna of Nebraska include *Paracoluber storeri* and the indeterminate "archaic" colubrid. It may be that these forms indicate that the Canadian locality is slightly older.

The Recent Fauna

The herpetofauna of today (Checklist) is an assemblage typical of semi-arid grasslands, similar to those in eastern Alberta, extreme western Manitoba, western North and South Dakota, northwestern Nebraska, northeastern Wyoming, and most of Montana. The checklist is modified from Logier and Toner (1961). These Recent forms reflect changes from the more moist, subtropical climate of late Miocene times to the semi-arid, north temperate climate of today. Generically, the only similarities between the Recent herpetofauna and the late Miocene herpetofauna are *Scaphiopus* (*Spea*) and *Bufo* which are found in both faunas. Other elements of the late Miocene fauna have their closest affinities with forms to the south.

Discussion and Summary

Two fossil deposits in Saskatchewan, one representing the early Oligocene and the other representing the late Miocene, give us the only glimpse of amphibian and reptilian life of the Tertiary of Canada. The late Oligocene Cypress Hills amphibian fauna is modern in that all four genera represented are living today. All of the amphibians represent forms with southern or southeastern affinities today. The reptilian fauna is archaic, with only four of 13 genera extant. The living genera, *Trionyx*, *Geochelone*, *Crocodylus*, and *Crotaphytus* all occur well to the south and southeast of the area today.

The late Miocene Wood Mountain fauna is much more modern than the early Oligocene one as both amphibian genera and six of seven reptiles identified are extant at the generic level. The amphibian genera *Scaphiopus* (*Spea*) and *Bufo* occur in the area today, but the affinities of most of the reptilian groups lie with more southern forms. Some of these southern forms have living relatives distributed either to the southwest or to the southeast of the fossil locality today.

The presence of crocodilians and tortoises of the genus *Geochelone* in both faunas indicates a subtropical climate in the area both during early Oligocene and late Miocene times. Southern Saskatchewan is thought to have been an ecotonal area between the Temperate Unit of the Arcto-Tertiary flora and the Neotropical flora in the early Tertiary, but the climate supposedly cooled off and it is thought that the flora withdrew southward in the Miocene. But the herpetofauna indicates a subtropical climate in late Miocene times, and it seems that the flora must have been more mesophytic and with more southern species than it has today.

The modern herpetofauna of Saskatchewan is typical of a northern, semi-arid, grassland region, and it has a rather restricted number of species because of its northern latitude. Many of the late Miocene forms withdrew to the south.

Estes (1970) has pointed out that the ancestors of many Recent United States species must have developed in Canada. This is because in the early Tertiary a temperate deciduous forest is thought to have been present in Canada at the time most of the United States had a Neotropical flora. Estes' hypothesis is substantiated by the fact that the earliest records of the following United States and Mexican genera are from the Tertiary of Saskatchewan: *Ambystoma* (early Oligocene); *Rhinophrynus* (early Oligocene); *Scaphiopus* subgenus *Scaphiopus* (early Oligocene); *Hyla* (early Oligocene); *Clemmys* (Paleocene); *Crotaphytus* (early Oligocene); *Ophisaurus* (late Miocene); and *Sceloporus* (late Miocene).

At present we have only a handful of specimens from Saskatchewan to depict the rich herpetological life that must have been present in Canada during the Tertiary. Not only are more specimens from the Cypress Hills and Wood Mountain formations desirable, but faunas are needed to fill in the gaps in the record. It would be especially exciting to get faunas from the interval between early Oligocene and late Miocene times and between late Miocene and Recent times, especially Pliocene faunas.

Literature Cited

AXELROD, D.I.
1958　Evolution of the Madro-Tertiary geoflora. Bot. Rev., 24: 433–509.

CHANEY, R.W.
1947　Tertiary centers and migration routes. Ecol. Mono., 17: 139–148.

COPE, E.D.
1891　On Vertebrata from the Tertiary and Cretaceous rocks of the North West Territory. 1. The species from the Oligocene or Lower Miocene beds of the Cypress Hills. Geol. Surv. Canada, Contrib. to Canadian Paleont., 3(1): 1–25.

ESTES, R.
1970　New fossil pelobatid frogs and a review of the genus *Eopelobates*. Bull. Mus. Comp. Zool., 139(6): 293–339.

ESTES, R. AND J.A. TIHEN
1964　Lower vertebrates from the Valentine Formation of Nebraska. American Midl. Nat., 72(2): 453–472.

HECHT, M.K.
1959　Amphibians and reptiles. *In* Paul O. McGrew, The geology and paleontology of the Elk Mountain and Tabernacle Butte area, Wyoming. Bull. American Mus. Nat. Hist., 117(3): 117–176.

HIBBARD, C.W.
1960　An interpretation of Pliocene and Pleistocene climates in North America. The President's Address. 62nd Ann. Rept. Michigan Acad. Sci., Arts., Letts.: 5–30.

HOLMAN, J.A.
1963　A new rhinophrynid frog from the early Oligocene of Canada. Copeia, 4: 706–708.
1968　Lower Oligocene amphibians from Saskatchewan. Quart. Jour. Florida Acad. Sci., 31(4): 273–289.
1970　Herpetofauna of the Wood Mountain Formation (Upper Miocene) of Saskatchewan. Canadian Jour. Earth Sci., 7(5): 1317–1325.
1971　Climatic significance of giant tortoises from the Wood Mountain Formation (Upper Miocene) of Saskatchewan. Canadian Jour. Earth Sci., 8(9): 1148–1151.
1972　Herpetofauna of the Calf Creek local fauna (Lower Oligocene: Cypress Hills Formation) of Saskatchewan. Canadian Jour. Earth Sci., 9(12): 1612–1631.

KENDEIGH, S.C.
1961　Animal ecology. Englewood Cliffs, New Jersey: Prentice-Hall. 468 pp.

LAMBE, L.M.
1906　Description of a new species of *Testudo* and *Baëna* with remarks on some Cretaceous forms. Ottawa Nat., 19(10): 187–196.
1908　The Vertebrata of the Oligocene of the Cypress Hills, Saskatchewan. Geol. Surv. Canada, Contrib. Canadian Paleont., 3(4): 1–64.

LOGIER, E.B.S. AND G.C. TONER
1961　Checklist of the amphibians and reptiles of Canada and Alaska. Roy. Ontario Mus. Contrib. 41, 92 pp.

RUSSELL, L.S.
1934a　Revision of Lower Oligocene vertebrate fauna of the Cypress Hills, Saskatchewan. Trans. Roy. Canadian Inst., 20(1): 49–67.
1934b　Fossil turtles from Saskatchewan and Alberta. Trans. Roy. Soc. Canada, 28: 101–110.

1936 New and interesting mammalian fossils from western Canada. Trans. Roy. Soc. Canada, 3rd ser., 30(4): 75–80.

1938 The skull of *Hemipsalodon grandis*, a giant Oligocene oreodont. Trans. Roy. Soc. Canada, 3rd ser., 32(4): 61–66.

1940 Titanotheres from the Lower Oligocene Cypress Hills Formation of Saskatchewan. Trans. Roy. Soc. Canada, 3rd ser., 34(4): 89–100.

1948 Geology of the southern part of the Cypress Hills, southwestern Saskatchewan. Dept. Min. Res., Geol. Surv. Canada, Rep. no. 8: 1–60.

WEIGEL, R.D.

1963 Oligocene birds from Saskatchewan. Quart. Jour. Florida Acad. Sci., 26(3): 257–262.

Hypsilophodont Dinosaurs:
A New Species and Comments
on Their Systematics

William J. Morris
Research Associate in Vertebrate Paleontology, Museum of
Natural History, Los Angeles County, and Professor of
Geology, Occidental College, Los Angeles

Abstract

The ornithischian archosaurs of the family Hypsilophodontidae
have been considered to be a cohesive group sharing a similar
morphology. Recent discoveries of additional material provisionally
assigned to *Thescelosaurus neglectus*, *?T. garbanii*, new species,
and *?T.* sp. indicate that a greater diversity existed than was
formerly realized. *Thescelosaurus* appears to have been a specialized late member of the hypsilophodont complex.

Classification of hypsilophodonts and other ornithischian archosaurs according to locomotor capabilities and presence or absence
of cheeks is discussed. Such classifications are considered premature
and not warranted by the specimens at hand. A return to a more
conservative classification is suggested.

Introduction

The earliest reported ornithischians are Upper Triassic members of the
Hypsilophodontidae. At least three different classifications are being advocated
for these archosaurs: those of Steel (1969), Thulborn (1971), and Galton
(1972). The classification favoured in this discussion is that of Thulborn (1971),
but with one difference. I do not believe that the pachycephalosaurs should be
included within the Hypsilophodontidae. This Upper Triassic through Cretaceous
array of small bipedal ornithischians has been termed the "Hypsilophodont
Plexus" by Thulborn (1971, p. 77), who states: "This succession of hypsilophodont dinosaurs ... might best be envisioned as an evolutionary plexus in
which each genus represents an individual strand. This concept of a plexus

(rather than a simple lineage) lacks the implication that any one genus might be directly related to another." As discussed later, I concur in this treatment as it offers a convenient device whereby a less than adequately known assemblage, fundamental to our understanding of ornithischian phylogeny, can be discussed without making premature taxonomic judgments.

Recent discoveries of two hypsilophodontids from the Hell Creek Formation of Montana (one from Lance Formation of South Dakota and a review of one from the Edmonton Formation of Alberta) provide additional information and suggest that greater diversity existed among this group than was previously recognized.

Thescelosaurus has been reported from the Upper Cretaceous of Niobrara County, Wyoming; Dawson County, Montana; Garfield County, Montana; Butte County, South Dakota; and Red Deer River, Alberta, Canada. Unfortunately those reported are incomplete or lack corresponding parts so that seldom can morphological characteristics be directly compared at the specific level. The holotype of *T. neglectus*, USNM 7757, lacks the skull, cervical vertebrae, humeri, scapulae, and coracoids. Its paratype, USNM 7759, does have a few cervicals, but it also lacks the skull. Referred specimens USNM 7760, 7761, 8065, 8016, and 5031 are even more fragmentary, and they lack skulls. Gilmore (1913) assigned all seven specimens to *Thescelosaurus*, but only USNM 7757 and 7758 were referred to *T. neglectus* Gilmore 1913. Sternberg (1940) reported and described *Thescelosaurus edmontonensis* (NMC 8537) from the Edmonton Formation of Alberta, Canada. Although the skull of this specimen is very fragmentary, Sternberg (1940) figured and described a frontal, postorbital and dentary. Parks (1926) described a small archosaur from the Upper Cretaceous Edmonton Formation of the Red Deer River, Alberta, as *Thescelosaurus warreni*, but Sternberg (1937) referred it to a distinct genus, *Parksosaurus*.

The tacit assumption has been made, both formally and informally, by Sternberg, that the skull of *Thescelosaurus* was probably similar in shape to that of *Parksosaurus warreni* (Parks, 1926; Sternberg, 1940). Galton (1973) comments, however, that the proportions of frontal to nasals are different in the two genera. In addition, the tarsal-metatarsal joints of *T. neglectus* and *T. edmontonensis* have not been closely compared. This is a significant joint, as the ankles of the two species appear to have been quite different. This difference became apparent after the recent discovery in Montana of a specimen (LACM 33543) described in this paper. Not only do the recently discovered specimens suggest differences in tarso-metatarsal morphology, but they also show that considerable variation existed in the skulls.

Notations are as follows: LACM: Museum of Natural History, Los Angeles County; SDSM: South Dakota School of Mines and Technology; USNM: National Museum of Natural History (United States National Museum); NMC: National Museums of Canada.

Systematics

Three previously unreported specimens form the basis of this consideration. One, LACM 33543, is referred to *Thescelosaurus neglectus*, while the other two are considered new species. Because of the fragmentary nature of the specimens previously referred to *Thescelosaurus*, most do not share features in common. As a result, when more complete specimens are discovered, many of the characteristics considered here to be of specific level may prove to be of generic significance.

Order Ornithischia Seeley, 1888
Suborder Ornithopoda Marsh, 1871
Family Hypsilophodontidae Dollo, 1882
Subfamily Thescelosaurinae Sternberg, 1940

Thescelosaurus neglectus Gilmore, 1913
Figs. 1, 2, 3D, 4

Referred Material
LACM 33543, consisting of skull fragments, vertebral column lacking atlas and caudals, numerous ribs, left and right ilia, ischia, and pubes, several metapodials and phalanges of the pes and manus. Only elements that provide additional information concerning this species will be described.

Locality and Age
LACM, locality *V* 3152, Hell Creek Formation, Lance, in the NE/4 of NW/2, SEC. 12, T21N, R42E, Garfield County, Montana.

Description

SKULL
The skull remains are very fragmentary and mostly disarticulated (Fig. 1). Identification of two right jugals indicates the presence of more than one individual in this locality. One individual is represented by much of the dorsal and nuchal parts of a cranium (Fig. 1A, C, D). In addition, a jugal and left dentary (Fig. 1B, E) have been referred to this individual on the basis of proportions and degree of weathering. It is not certain, because of differences in degrees of weathering, if the right dentary belongs to this or another individual (Fig. 1, E2, E3).

The only other reported skull material of *Thescelosaurus* was referred to *T. edmontonensis* (NMC 8537) by Sternberg (1940). However, the tarso-metatarsal joint of *T. edmontonensis* is significantly different from that of *T. neglectus*. Should more complete specimens confirm this observation then NMC 8537 may well be assigned to another genus. Then LACM 33543 would be the only cranial material that could definitely be referred to *Thescelosaurus*.

The occipital of LACM 33543 is decidedly massive, contrasting markedly with the more delicately proportioned bones of the cranial roof. The massive occipital condyle is formed by a large ventrally directed basioccipital, and the two exoccipitals. The exoccipital and basioccipital are not ankylosed but have well-defined sutural contacts (Fig. 1, c1,c2). The entire region, including the opisthotic, is more massive than that found in either *Parksosaurus* or *Stegoceras*.

Sternberg (1940) states that a generic characteristic of *Thescelosaurus* is the presence of parallel ridges and grooves trending dorsoventrally along the outer face of the cheek-teeth. This characteristic is variable in LACM 33543 as some ridges and grooves are parallel, while others are not.

Sternberg (1940, p. 482) reports the presence of 16 dentary teeth along a distance of 78 mm in *T. edmontonensis*. The LACM specimens have a preserved alveolar margin of about this length, however the minimum tooth count is either 18 or 19. Many of their alveoli are double, particularly the more anterior ones, suggesting the presence of one replacement tooth erupting on the buccal side of its more worn precursor.

VERTEBRAL COLUMN

Gilmore (1915, p. 592) reported 16 dorsal and five sacral vertebrae for *T. neglectus*. Sternberg (1940, p. 485) disagreed with this count. He suggested that the anterior presacral reported by Gilmore was probably the last cervical, and that the last dorsal is actually the first sacral. Therefore, his count was 14 dorsal but six sacral. Steel (1969, p. 9) states that the vertebral count for the Hypsilophodontidae, to which *Thescelosaurus* was assigned, is "approximately 25 presacral vertebrae, including 9 cervicals and 4–6 sacrals." LACM 33543 has an almost complete sacral and presacral vertebral column, as only the atlas is missing. There are 27 presacral vertebrae distributed as follows: axis, six other cervicals, 20 dorsals, and five sacrals. The total revised presacral vertebrae count is therefore 28.

LACM 33543 differs from the description of the type of *T. neglectus* in that only the first and second dorsal vertebrae have the dorsally angled transverse processes reported by Gilmore (1913, 1915). The rest of the dorsals have laterally directed transverse processes.

The anterior margins of the neural spines of the cervical vertebrae of the previously reported specimens of *Thescelosaurus* apparently were not preserved and have not been figured. The anterior margins of the neural spines of LACM 33543 are very thin, almost transparent, blades (Fig. 2B, C, D). These thin blades are rugose and serrate along the anterior and part of the dorsal margins. Similar features have not, to my knowledge, been reported for any other archosaur.

Fig. 1 *Thescelosaurus neglectus* (LACM 33543), referred specimen.
  A. Lateral aspect of skull as restored.
  B1. External view right jugal. B2. Internal view right jugal.
  C1. Right lateral aspect of occipital region. C2. Posterior view of occipital region.
  D. Dorsal aspect of cranium.
  E1. Lateral aspect of left dentary. E2. Dorsal aspect of composite lower jaw. Right dentary many not belong to same individual as rest of elements. E3. Interior view of right dentary.
  Abbreviations: bo, basioccipital; pa, parietal; fr, frontal; ju, jugal; de, dentary; po, articulation of postorbital; la?, articulation for lacrimal; qa, articulation for quadrate; op, opisthotic; eo, exoccipital; mc, groove for Meckel's cartilage; sp, groove and articulation for splenial.

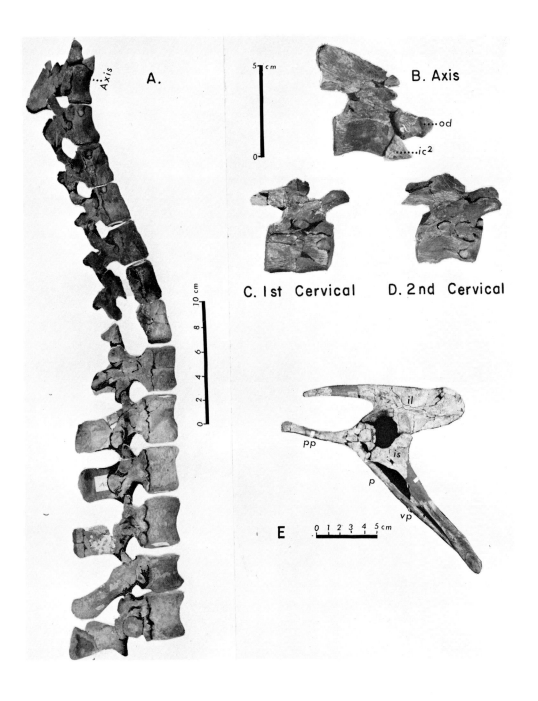

A.

B. Axis

...od

......ic²

C. 1st Cervical   D. 2nd Cervical

E

braces, a point that has not been previously stressed. The curvature of the distal part of the shaft of the pubis, and the geometry of the ischiadic peduncle and the ventral process of the ischium are such that when these two elements are properly articulated they firmly interlock. The entire structure when lifted by the prepubic process holds tightly together. It would appear that the structure is designed to resist anterior rotation of the ischium and pubis during retraction of the femur when the M. adductoris is contracted. During protraction of the femur, contraction of the Mm. pubotibialis and, perhaps, ambiens would tend to rotate the pubis anteriorly in a direction opposite of that during retraction. The buttressing of the ischiadic peduncle and the ventral process of the ischium would function to counteract this stress. However, the tendency for resisting an anterior rotational force would appear to have placed considerable stress on the relatively weak bony juncture between the iliac peduncle and the main body of the pubis. If this analysis is correct, one can only speculate that strong ligamental or cartilagenous supports existed which served to buttress the prepubis against the vertebral column or the transversely triangular anterior ramus of the ilium.

Iguanodon appears to have had similar structures to counter rotation of the pelvis but lacks the ventral (obturator) process of the ischium. Instead, forward rotation of the ischium is resisted by a more proximal buttress closely apposed to the medial section of the pubis and a strong ischio-pubic contact. The pelvis of *Parasaurolophus* appears to have a third type of construction designed to counteract similar rotational stresses. It has an amazingly strong interlocking articulation between the ischium and ilium as well as a peculiar ischial flange extending ventrally along the anterior part of the acetabulum and lying against the prepubis. Interlocking or counter-rotational structures appear to be absent in the other hadrosaurs, and particularly in the crested forms. In my opinion the presence of counter-rotational structures in *Thescelosaurus, Iguanodon,* and *Parasaurolophus* indicates a more generalized and terrestrial habit for these forms, while the lack of such devices, particularly in the crested hadrosaurs, is indicative of a more specialized and aquatic habit. Severe stresses resulting from body weight and locomotory thrust would not be present in an aquatic habitat.

?*Thescelosaurus garbanii* sp. nov.
Figs. 3A–C, 5D–F

Holotype
LACM 33542.

Horizon
Hell Creek Formation, Upper Cretaceous.

Locality
LACM V3152; T.21N, R42E, NE/2, NW/4, Sec. 22, Garfield County, Montana.

Table 1. Comparative measurements of maximum length of elements of the pes. (All measurements in mm; from Parks, 1926, except for LACM 33543 and LACM 33542.

Element	Parksosaurus warreni holotype	Thescelosaurus neglectus, holotype USNM 7757	Hypsilophodon foxii, holotype	Thescelosaurus (?) garbanii (sp. nov.) LACM 33542	Thescelosaurus (?) neglectus LACM 33543
Metatarsal I	88	65	—	92	
II	135	112	93	151	
III	151	127	105	186	
IV	119	106	88	137	
V	35	28	32	46	
Phalanx I^1	53	51	—	67	
I^2	51	53	17	—	
II^1	52	53	23	70	
II^2	35	34	14	55	32
II^3	61	47	21	65	42
III^1	51	46	25	68	—
III^2	38	38	17	59	32
III^3	37	35	13.5	49	34
III^4	64	54	23.5	72	—
IV^1	43	35	17	49	—
IV^2	34	26	12	44	—
IV^3	31	22	8	34	—
IV^4	32	19	8	29	—
IV^5	42	40	17	61	33
Digit I	192	169	—	203	
II	283	246	150	322	
III	338	300	184	415	

Table 2. Ratio of tibial to digital lengths.

Digit	Parksosaurus warreni	Thescelosaurus neglectus USNM 7757	Hypsilophodon foxii	T.(?) garbanii LACM 33542
I	1.7:1	1.8:1	—	2.0:1.0
II	1.0:1.0	1.2:1.0	1.5:1.0	1.3:1.0
III	1.0:1.0	1.0:1.0	2.2:1.0	1.0:1.0
IV	1.1:1.0	1.2:1.0	1.5:1.0	1.2:1.0

Material

Holotype consisting of five posterior cervical and 11 anterior dorsal vertebrae; left pes, tarsus, tibia, fibula, and distal part of femur.

Etymology

Named in honour of its discoverer, Mr. Harli Garbani.

Diagnosis of Species

?Thescelosaurus garbanii is about a third larger than described specimens of *T. neglectus* and *Parksosaurus* or nearly twice as large as *Hypsilophodon* (Table 1), and possesses an unique ankle. The fibula is shorter than the tibia. The distal terminus of the fibula is very slender and closely appressed to the lateral side of the tibia. The fibula is articulated only loosely with the calcaneum. The calcaneum is characteristic of the species in that it consists of a thin plate fitting into a shallow depression on the lateral side of the distal end of the tibia. The astragalus is a wide element appressed to and covering the distal end of the tibia. The astragalus is firmly united to the tibia. The astragalus forms the proximal part of the hinge between the proximal and distal tarsal elements. Significantly, the calcaneum does not extend ventrally to the ventral margin of the astragalus and, therefore, does not form a part of the tarso-metatarsal hinge as it does in *T. neglectus* (Fig. 3D). Aside from this unique construction of the tarso-metatarsal hinge, the pes corresponds closely in form and proportion to that of *T. neglectus*.

The dorsal and cervical vertebrae, although larger than those reported for *T. neglectus*, have the same configuration and rugosities over the lateral and ventral parts of the centrum.

Discussion

The assignment of LACM 33542 to the genus *Thescelosaurus* would be questionable if it were not that the same arrangement of the calcaneum and astragalus is probably present in *T. edmontonensis*. A comparison of the tarso-metatarsal hinge in *T. edmontonensis* Sternberg 1940 (Fig. 3E) with that of *?T. garbanii* (Fig. 3A, B) shows significant similarities in the location and width of the astragalus which forms the articulation for the entire proximal side of the tarso-metatarsal hinge. The calcaneum of *T. edmontonensis* was not described, but there is clear indica-

Fig. 3 A, B, C. *?Thescelosaurus garbanii*, sp. nov.
   A. Lateral view of left pes.
   B. Palmar view of left pes.
   C. Dorsal view of left pes.
   D. Dorsal view of pes of *T. neglectus* (holotype; from Gilmore, 1915), compare geometry of calcaneum and astragalus with A, B, and C.
   E. Tibia and fibula of *T. edmontonensis* (holotype; from Sternberg, 1940). Note, a, facet for calcaneum and geometry of astragalus. Compare with A, B, C, and D.
   F. Femur, tibia and fibula of *Stegoceras* (from Gilmore, 1924). Note distal end of tibia and facet for astragalus are very similar to those in A, B, C, and E.
   Abbreviations: Fe, distal end of femur; T, tibia; F, fibula; Ca, calcaneum; As, astragalus; I–V, 1st through 5th digits.

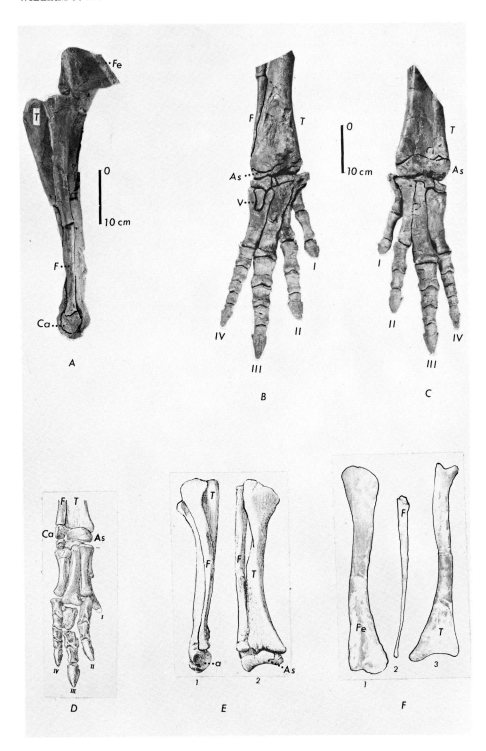

tion of a shallow pit on the lateral surface of the tibia into which the calcaneum would have fitted. This arrangement is very similar to that of *?T. garbanii.*

Gilmore (1924) figured the tibia and fibula of the pachycephalosaurid *Stego-ceras* (Fig. 3F). Unfortunately the astragalus and calcaneum were not reported and the fibula is poorly preserved. However, the slenderness of the distal end of the fibula and the relatively broad tibia suggest that the proximal tarso-meta-tarsal joint is very like that of *?T. garbanii.* H. Osmolska (1973, pers. comm.), after correspondence concerning this similarity between *Stegoceras* and *?T. gar-banii,* sent me photographs of the Mongolian pachycephalosaurid *Homalocephale calathoceras.* Although the skull is very similar to *Stegoceras,* the excellent photo-graphs show that the proximal tarso-metatarsal joint of the pes is not at all like that of *?T. garbanii* or, for that matter, *T. edmontonensis.* Furthermore the pelvis of *H. calathoceras* is very different from that described for *Thescelosaurus.*

The pes of *?T. garbanii* was found articulated (Fig. 3B, C). In Fig. 3D, E, F, the elements have been oriented, as indicated by the curvature of the particular facets, to portray positions during maximum protraction and retraction. The geometrical possibilities suggest that *?T. garbanii* may have been digitigrade during the power stroke with the phalanges pressed against the ground and the metatarsus nearly vertical. I suggest that a similar situation occurs in the power stroke of both the emu and cassowary. The firm union of the astragalus and tibia in *?T. garbanii* sug-gests a modification, similar to that in carnosaurs, in which stress is transmitted from the lower foot to the tibia by the broadly expanded astragalus. This modifi-

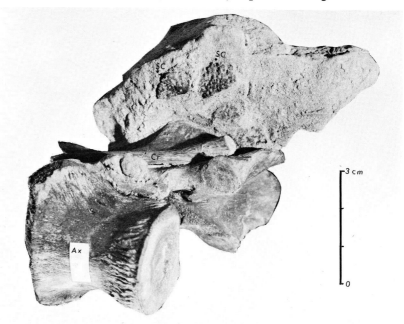

Fig. 4 Axis of *T. neglectus* (LACM 33543). Note preservation of two well-preserved dorsal scutes (SC) and one poorly preserved scute in matrix. Specimen figured as collected. Ax, centrum of axis; Cr, cervical rib.

cation of the proximal tarsal joint suggests that ?*T. garbanii* was a cursorial form and probably bipedal.

Recent reconstructions of hypsilophodonts and pachycephalosaurids (see, for example, Galton, 1970, and Thulburn, 1971) indicate very agile creatures, nimbly balanced on the slender tips of their toes. This may well be in error and instead the locomotive thrust may have been attained from a digitigrade stance with all but the proximal phalanx pressed against the substratum.

?*Thescelosaurus* sp.

Fig. 6

Specimen

SDSM 7210.

Horizon

Hell Creek Formation, Upper Cretaceous.

Locality

Hardin County, South Dakota (additional data lacking).

Material

Left half of cranium; left dentary and fragment of right dentary; centra of two dorsal vertebrae; two dorsal vertebrae lacking transverse process and neural spines; two manual phalanges.

Description of Specimen

In this small archosaur the skull is approximately 20 cm long and 15 cm high, as restored (Fig. 5A, B), and the width of its orbit is approximately one-third of the skull length. A small antorbital fenestra is present and bounded posteriorly by the lachrymal and maxilla and ventrally by the maxilla. The maxillary tooth-row is deeply recessed towards the midline, and situated beneath a prominent overhanging ridge along the ventral part of the maxilla. This ridge is marked by anastomosing rugosities. Eleven maxillary teeth are present, each of which bears a vertical ridge margined by finer costations along the centre of its lateral surface. Some of the costations parallel the central ridge, while others are arranged in a fan-like configuration. The dentary teeth are very similar, but differ in having an outer rather than inner wear-facet. The wear-facets indicate a well-developed shearing function. Each maxillary and dentary tooth has a single replacing counterpart. There is no evidence to indicate more than these two tooth-rows.

Teeth are present in the premaxilla. Presence or absence of teeth in the predentary is not known for this bone is not preserved. The premaxillary teeth are simple, recurved cones. There are two erupted, functional premaxillary teeth but only the most anterior of these is completely preserved. The other premaxillary teeth had not erupted. These unerupted premaxillary teeth are more conical and only slightly smaller than their erupted counterparts. The outer surfaces are characterized by very fine striations converging towards the apex of the tooth. As with the dentary and maxillary teeth there is evidence for only one row of replacing teeth. The apex of the one erupted and complete premaxillary tooth appears worn. Whether this resulted from apposition to a bill on the predentary or from occlusion with predentary teeth cannot be determined from this specimen.

A well-developed supraorbital extends two-thirds the distance across the dorsal margin of the large orbit, much as in *Heterodontosaurus*.

The lower jaw is more massive than that of *T. neglectus*, and the width of the ramus is almost twice that in *T. neglectus*. The lower jaw also differs from *T. neglectus* in that the splenial is not visible in lateral view. In addition a prominent foramen is located midway along the suture between the dentary and surangular.

The centra of the dorsal vertebrae (Fig. 5c) do not differ significantly from those described for *T. neglectus*. As in the other species of *Thescelosaurus* the lateral and ventral surfaces of the centra are extremely rugose suggesting attachment for powerful lateral musculature.

Discussion

The skull has two very characteristic morphological features that appear to be of a generalized nature. These are the presence of premaxillary teeth which differ from those of the maxillary. Similar tooth differentiation has been reported for the Heterodontosauridae (*sensu* Kuhn, 1966) and for the Hypsilophodontidae (*sensu* Dollo, 1882). ?*T.* sp. does not appear to have had enlarged lateral premaxillary teeth, which are characteristic of *Heterodontosaurus*. However, the premaxillary of the only known specimen of the latter genus is very poorly preserved.

Galton (1972) has suggested that certain of the ornithopods, including *Thescelosaurus*, had cheeks derived from the trigeminal musculature. He suggests that the recessed dentition is morphological evidence for this condition. In *Thescelosaurus* the presence of cheeks would have to be deduced from the dentary as, prior to the discovery of this specimen, the shape of the maxilla could only be inferred by comparisons with the related form *Parksosaurus*. The deeply recessed maxillary dentition and the lateral rugosities along the posterior margin of the maxillary, and continuing onto the jugal, strongly suggest that Galton's interpretation is correct. The condition is, however, much more pronounced than in any other described ornithopod.

A first impression is that ?*T.* sp. is a very specialized hypsilophodontid by reason of the deeply recessed maxillary tooth-row, laterally flaring premaxilla,

Fig. 5 A, B, C. ?*Thescelosaurus* sp. (SDSM 7210).
    A. Stereophotograph of skull in left lateral aspect.
    B. Stereophotograph of anterior view of premaxillary.
    C. Lateral and ventral aspect of dorsal vertebrae.
    D, E, and F. Left pes of ?*Thescelosaurus garbanii*.
    D. Digits in protracted position with digits I, II, and IV oriented in position relative to illustrated angle of metatarsus.
    E. Digits of left pes arranged to illustrate positions during retraction.
    F. Dorsal perspective view of left pes with digits oriented in digitigrade position as collected.

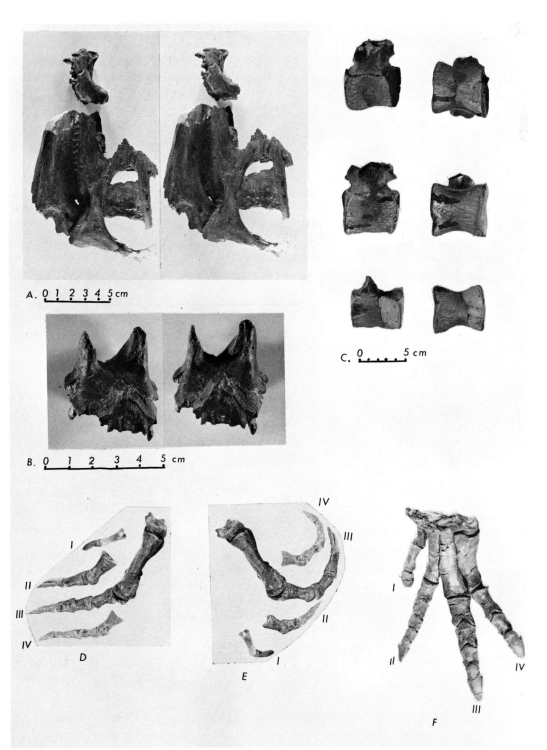

A. 0 1 2 3 4 5 cm

B. 0 1 2 3 4 5 cm

C. 0   5 cm

D

E

F

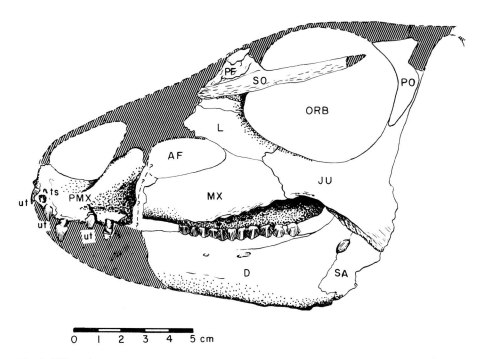

Fig. 6 ?*Thescelosaurus* sp. (SDSM 7210). Restored portions are shaded and articulation between maxillary and premaxillary shown in approximate position as this area was not preserved. Abbreviations: AF, anterior formen; D, dentary; JU, jugal; L, lacrimal; MX, maxillary; ORB, orbit; PF, prefrontal; PMX, premaxillary; PO, postorbital; SA, surangular; SO, supraorbital; ts, probable alveolus of unerupted premaxillary tooth; ut, unerupted premaxillary teeth.

recumbent premaxillary teeth, and superorbital. Recumbent premaxillary teeth are characteristic of the Triassic heterodontosaurids, and a supraorbital bone has been reported in other members of the Hypsilophodontidae (*Heterodontosaurus*, *Dysalotosaurus*) and in the Iguanodontidae (*Camptosaurus*). However, these features may be characteristic of all species of *Thescelosaurus* as in species other than ?*T*. sp. the lateral and ventral portions of the skull are unknown. If this is the case, then assumptions that *Thescelosaurus* and *Parksosaurus* have morphologically similar skulls would be quite unwarranted. Indeed the discovery of ?*T*. sp. leads to the strong inference that only *Parksosaurus*, *Laosaurus*, and perhaps *Hypsilophodon* maintained the generalized ornithopod skull and that *Thescelosaurus* is a very specialized lineage.

Comments on Graviportal and Cursorial Dinosaurs

Recently two revisions of ornithopod archosaurs have been proposed; these attempts were made to reconcile taxonomic differences between those small forms included in the Hypsilophodontidae, *sensu* Dollo (1882) and Steel (1969). Thulborn (1971) based his classification upon osteological morphology, and, with the exception of including the Pachycephalosauridae within Hypsilophodontidae, his classification does not differ significantly from those proposed by earlier authors. Thulborn infers that the various lineages are phyletically related in a complex way that is difficult to characterize. Thus he uses the phrase "Hypsilophodont plexus" to describe, informally, the relationship between the various genera, from *Fabrosaurus* to *Thescelosaurus*, which may be thought of as more generalized ornithopods from which the iguanodonts, ceratopsians, hadrosaurs and others were derived. Certainly the fabrosaurs are structurally and temporally suitable ancestors for later ornithopods. However, as indicated in this paper, *Hypsilophodon* and *Thescelosaurus*, like the heterodontosaurs, are far too specialized to be included in basic ornithopod stock. These three forms exhibit certain very similar features in the structure of the pelvis, although their cranial and dental morphology appears to have been very different. I cannot perceive any characteristics, excepting some very superficial ones, which are shared by heterodontosaurs, *Hypsilophodon* and *Thescelosaurus*, on the one hand, and the ceratopsians, hadrosaurs, and iguanodonts on the other. In addition I surmise that Thulborn's difficulty in taxonomic resolution at the family level is an artifact of poorly preserved or incomplete material.

Galton (1972) has revised the familial classification of the primitive ornithopods based upon the presence of cheeks and the locomotor capabilities of these small dinosaurs. It is my opinion that his proposed classification is not valid for the following reasons: (1) the paucity of cranial material in some genera makes the presence of cheeks an assumption and not a demonstrable, morphological fact; (2) the assessment of locomotor capabilities depends upon more than comparative ratios of rear limbs; (3) palaeontological classifications should not be based upon functional behaviour, deduced from partial skeletons, having few if any modern analogues; and (4) locomotor capabilities have not been demonstrated to be of phylogenetic and therefore of taxonomic significance.

Galton (1972) removes the ornithischians *Fabrosaurus* and *Echinosaurus* from the Hypsilophodontidae and erects the family Fabrosauridae for them. He places *Thescelosaurus* in the Iguanodontidae, a conclusion based on limb and body proportions and one with which I disagree. He states that "*Fabrosaurus* and *Echinodon* are the only ornithischians resembling living reptiles in having marginally positioned teeth. ... I correlate marginal teeth with the absence of cheeks, and an inset position with the presence of cheeks." Unfortunately this can only be demonstrated for ?*Thescelosaurus* sp. (SDSM 7210), and not for the type-species *T. neglectus*, a critical species in the phylogeny, as its holotype and paratypes consist only of postcranial material.

A further difficulty concerns *T. edmontonensis* described and named by Sternberg (1940). Although the cheek region is missing from the skull the mandibular ramus was collected. Even if one were to assume that the presence or absence of cheeks could be validly deduced from the lower jaw, there is some question as regards the taxonomic assignment of *T. edmontonensis*. Sternberg's plates, reproduced here as Fig. 3E, show that the astragalus of *T. edmontonensis* covers almost the total distal surface of the tibia and that the calcaneum, although not preserved, appears to have occupied a deep pit on the distal, external side of the tibia. The fibula narrows distally and does not extend ventrally as far as the astragalar-tibial facet. The reduction of the fibula distally, the position of the calcaneum, and the astragalus forming almost the entire proximal hinge-joint of the foot are radically different from conditions in the rear limb of *T. neglectus*. In *T. neglectus* the proximal hinge-joint operated between the tarsus and relatively unreduced astragalus-calcaneum. This major difference in foot morphology strongly suggests that *T. edmontonensis* can only be provisionally assigned to *Thescelosaurus*. I believe that these morphological differences make questionable the assumption that *T. neglectus* had a dentary like that of *T. edmontonensis*. Description in this paper of *?T. garbanii* sp. nov., from the Hell Creek beds of Montana, further substantiates that different foot constructions were present in forms now included within *Thescelosaurus*.

I also question Galton's assumption that cheeks were a necessary adaptive feature for the success of these small, terrestrial, herbivorous ornithopods. Many genera of modern iguanid lizards are well adapted to a herbivorous habit without possessing cheeks. Galton suggests that a tooth row inset buccally would indicate cheeks, but inspection shows that the tooth row of *T. edmontonensis* is not inset proportionately more than that of *Iguana*.

Galton considers both *Pachycephalosaurus* and *Stegoceras* as graviportal forms. However, these two genera were included in the Pachycephalosauridae not on the basis of locomotor habit, but on the morphological similarity of the crania (Sternberg, 1933; Brown and Schlaikjer, 1943). In point of fact the postcranial skeleton of *Pachycephalosaurus* has not been reported, so it is difficult to follow Galton's conclusion that this animal was indeed graviportal.

From references cited Galton apparently follows Gregory's concept of graviportal. To my knowledge "graviportal" was first used by Gregory (1912) in an article about quadrupedal locomotion in hoofed animals. He states: "This word [graviportal] has been invented by Professor Osborn to describe the conditions in heavy-bodied animals with long proximal and short distal limb segments." Clearly Osborn had two criteria in mind and graviportal locomotion was a function of weight as well as limb proportions. Gregory further expanded on these two basic qualifications and not only discussed length of limb as important to cursorial and graviportal habit but also "angle of stride" and "acceleration increment of stride due to ballistic power of limbs." While "angle of stride" is apparent, what Gregory meant by "ballistic power" needs some explanation. To quote Gregory, "This ballistic power may be defined as excess propulsive power over and above that which is necessary to move the limbs as stilts and to support the weight of the body; it is expended in lengthening the stride."

Galton includes, among others, *Thescelosaurus* with *Kritosaurus* as graviportal forms. *Thescelosaurus* probably did not exceed 2.5 m in length and was lightly built, but *Kritosaurus* was relatively massively constructed and about 9.0 m long (Lull and Wright, 1942). Although both have similar proportions of the hind limbs, the weight of *Thescelosaurus* would not be sufficient to meet Osborn's criterion. In addition, the "acceleration and ballistic power" was not analyzed by Galton and it is indeed problematical that such could ever be correctly deduced from these ornithopod skeletons. Division into graviportal and cursorial forms on limb proportions, as Galton has done, does not appear justified on the basis of his reported evidence. Indeed, using limb proportions alone, one would conclude that the mole be classed with the horse.

Jenkins (1971) has presented a study of limb posture and locomotion in animals. He states: "... species that share similar limb posture and excursion patterns [e.g., the tree shrew and opossum] may manifest profound differences, not indicated by biomechanical data, in, for example, agility, duration or intensity of locomotor activity. Although mammalian locomotion and posture have been treated traditionally in biomechanical terms to the neglect of behavioural and physiological considerations, the latter should be accorded an equal value in any system of locomotory classification." I believe the same must apply to reptiles. Gregory and Osborn apparently recognized this complication in citing as end members within the spectrum of locomotive behaviour the horse and the elephant.

I do agree with Galton that the family Pachycephalosauridae should be retained and not placed within the Hypsilophdontidae as Thulborn (1971) suggests. However, my judgment rests not on assumed locomotory behaviour but on certain unique morphological characteristics of pelvis and tibia, as shown by *Stegoceras* and recently discovered stegocerid specimens from Mongolia (Osmolska, pers. comm., 1973).

Most paleontological classification should be based upon morphological homologies and not upon behavioural characteristics. While not a hard and fast rule, this certainly does apply within the group under discussion, as material is fragmentary and modern analogues are not available. For example, the discovery of two types of proximal hinge-joints in the new *Thescelosaurus* specimens indicates that two different habits of locomotion were obtained within the taxon. However, this inference would not be apparent from linear measurements of limb elements alone.

Galton (1972) removes *Thescelosaurus* from the Hypsilophodontidae and places it in the Iguanodontidae as he considers it, like *Iguanodon*, to have been "graviportal." The basis for this appears to be limb, and limb to body, proportions; however, these two genera differ in almost every other significant morphological structure. The morphologies of the pelves, pedes, manus, and skulls possess very little in common aside from fundamental reptilian features. They may, indeed, be related at a superfamilial level, but evidence at hand suggests that both *Thescelosaurus* and *Iguanodon* are much too specialized and morphologically dissimilar to be included in the same family.

Galton's classification, since it is based upon behavioral characteristics, can, at best, be a conceptual model for the definition of higher taxonomic categories, but it depends upon a higher level of abstract interpretation than can reasonably be inferred from the materials at hand.

In view of the confusion that has resulted from poorly defined terminology and lack of more complete material, I suggest that a conservative phylogeny would best serve our present state of knowledge. As a result I recognize the family Hypsilophodontidae to include: *Hypsilophodon, Dysalotosaurus, Laosaurus, Nannosaurus, Parksosaurus,* and *Thescelosaurus,* and the more primitive *Fabrosaurus* and *Echinodon.* This does not differ from the scheme proposed by Thulborn (1971), except that the pachycephalosaurids are retained as a family, *sensu* Sternberg (1945). It may well be that there is more than one genus now included in *Thescelosaurus.* Available specimens, however, are so fragmentary that this step is not warranted.

Summary

Discoveries of specimens from the Hell Creek Formation, Upper Cretaceous, of Montana and South Dakota, have significantly increased our knowledge of and existing information on the small bipedal dinosaur *Thescelosaurus*. A referred specimen of *T. neglectus* is described and compared with the holotype. Two new species provisionally assigned to *Thescelosaurus* are reported, ?*T. garbanii* and ?*T.* sp. Both of these have features indicating that *Thescelosaurus* is more specialized and more morphologically diverse than indicated by recently proposed phylogenies. In view of the incompleteness of reported material a more conservative classification than that recently proposed by Galton (1972) is suggested. *Thescelosaurus* is retained within the family Hypsilophodontidae rather than placed in the Iguanodontidae as Galton suggests. A critique of Galton's classification of ornithopods which is based upon presence of cheeks and locomotor capacities is presented and dismissed, as being premature and beyond our present knowledge of these dinosaurs.

Acknowledgments

Mr. and Mrs. William T. Sesnon, Jr., sponsored the project leading to the discovery of the Hell Creek specimens. Mr. Harli Garbani not only discovered these specimens but very ably prepared them. My thanks are extended to Dr. Robert Wilson, South Dakota School of Mines and Technology, for allowing us to examine the specimens from the Lance Formation. Mr. Eric Lichtward assisted in various comparative studies. My thanks are also extended to Messrs Michael Greenwald and Ralph Molnar for valuable discussions on many major points presented in this paper. Dr. Dale Russell, National Museums of Canada, commented on the several drafts of this paper, and his helpful criticisms and remarks are appreciated and gratefully acknowledged. A special note of thanks to Dr. H. Osmolska for providing photographs and information on Mongolian pachycephalosaurids.

Literature Cited

BROWN, B. AND E.M. SCHLAIKJER
   1943   A study of the troodont dinosaurs with the description of a new genus and four new
           species. Bull. Amer. Mus. Nat. Hist., 82: 115–150.

DOLLO, L.
   1888   Iguanodontidae et comptonotidae. C.R. Acad. Sci. Paris, 106: 775–777.

GALTON, P.
   1970   Pachycephalosaurid-dinosaur battering rams. Discovery, 6(1): 23–32.
   1972   Classification and evolution of ornithopod dinosaurs. Nature, 239: 464–466.

1973   Redescription of the skull and mandible of *Parksosaurus* from the Late Cretaceous with comments on the Family Hypsilophodontidae (Ornithischia). Roy. Ont. Mus. Life Sciences Contr., 89: 1–21.

GILMORE, C.L.
1913   A new dinosaur from the Lance Formation of Wyoming. Smith. Misc. Coll., 61: 1–5.
1915   Osteology of *Thescelosaurus*, an orthopodous dinosaur from the Lance Formation of Wyoming. Proc. U.S. Natl. Mus., 49: 591–616.
1924   On *Troödon validus*. Univ. Alberta, Dept. Geol. Bull., 1: 1–43.

GREGORY, W.K.
1912   Notes on the principles of quadrupedal locomotion and on the mechanism of the limbs in hoofed animals. Ann. New York Acad. Sci., 22: 267–294.

HUXLEY, T.H.
1870   On *Hypsilophodon foxii*. Quart. Jour. Geol. Soc. London, 25: 3–12.

JENKINS, F.A.
1971   Limb posture and locomotion in the Virginia opossum (*Didelphis marsupialis*) and in other non-cursorial mammals. Jour. Zool., 109: 303–315.

KUHN, O.
1966   Saurischia (supplementum I): Fossilium catalogus; I, Animalia (F. Westphal. ed.), The Hague, pars. 109, pp. 94.

LULL, R.S., AND W.E. WRIGHT
1942   Hadrosaurian dinosaurs of North America. Geol. Soc. Amer., Sp. Paper 40: 1–242.

PARKS, W.A.
1926   A new species of ornithopodous dinosaur from the Edmonton Formation of Alberta. Univ. Toronto, 21: 1–42.

STEEL, R.
1969   Ornithischia. Encyclopedia of paleoherpetology, Oskar Kuhn, ed., Gustav Fischer Verlag, Stuttgart, pt. 15: 1–85.

STERNBERG, C.M.
1933   Relationships and habitat of *Troödon* and the *Nodosaurus*. Ann. Mag. Nat. Hist., ser. 10, 2.
1937   Geol. Soc. Amer., Proc. for 1936, p. 375 (abstract).
1940   *Thescelosaurus edmontonensis*, n. sp., and classification of the Hypsilophodontidae. Jour. Paleo., 14(5): 481–494.
1945   Pachycephalosauridae, proposed for domeheaded dinosaurs, *Stegoceras lambei* n. sp. described. Jour. Paleo., 19(E).

THULBORN, R.
1971   Origins and evolution of ornithischian dinosaurs. Nature, 234(5234): 75–78.

A Late Cretaceous Vertebrate Fauna from the St. Mary River Formation in Western Canada

Wann Langston, Jr.
Texas Memorial Museum and Department of Geological
Sciences, University of Texas, Austin

Abstract

A limited exposure of late Cretaceous fresh-water sediments, known as Scabby Butte, in southern Alberta was among the first-discovered dinosaur localities in western Canada. Recent field investigations have greatly enlarged the known vertebrate fauna, of which the large "battering ram" ceratopsian, *Pachyrhinosaurus canadensis* Sternberg, is the most conspicuous element. The vertebrate and invertebrate fauna, palynology, and inferred depositional environments show that whereas the Scabby Butte sediments correlate in time with the lower part of the Edmonton (= Horseshoe Canyon) Formation to the north, the available habitats may have been closer to the sea and more marshy than the situations frequented by the relatively varied dinosaurian faunas of the classic Oldman and Edmonton deposits.

Although more limited in scope, the Scabby Butte assemblage shows close affinities with other well-known late Cretaceous faunas of western North America. It is deficient in the number and variety of the common Oldman and Edmonton dinosaurs; microvertebrates are locally abundant. No undescribed forms are recognized at Scabby Butte, but of 26 taxa identified at the generic level, 11 are new to the St. Mary River Formation.

Introduction

Among the earliest discoveries of dinosaur remains in western Canada were a few bones collected at an isolated patch of badlands indelicately called the Scabby

Butte, about 17 miles north-northwest of Lethbridge, Alberta. The fossiliferous deposits belong to the Upper Cretaceous St. Mary River Formation, which is perhaps best known to vertebrate paleontologists as the principal source of the unique "battering-ram" ceratopsian *Pachyrhinosaurus*. Fragmentary skulls of this bizarre and enigmatic dinosaur had been found at Scabby Butte in 1945, suggesting that further careful exploration there might prove rewarding.

With this prospect in mind, Dr. L.S. Russell and I (at the time, both on the staff of the National Museum of Canada) visited Scabby Butte in August 1955. Almost at once our discovery of two *Pachyrhinosaurus* skulls and numerous hadrosaurian bones revealed the existence of a bone-bed of a kind frequently encountered in the dinosaur fields of western Canada. This site was excavated in 1957, and, from a quarry some 20 by 50 feet (6 by 15 m) in diameter, were recovered more than 200 bones of ceratopsian and hadrosaurian dinosaurs. Exploration at Scabby Butte has continued intermittently since 1957, and several additional sites, including a second bone-bed of another sort, have been located and excavated. A varied fauna comprising more than 30 taxa of fish, amphibians, reptiles, and mammals has now been recovered. Limitations of space preclude systematic treatment of this material in the present volume; it will be published elsewhere (Langston, in press). In the meantime, a brief account of the occurrence and a summary of the fauna may be helpful to others interested in Scabby Butte and similar localities.

Specimen numbers referred to in this paper, preceded by the initials WL, are my field numbers for the Scabby Butte collection. The corresponding catalogue numbers in the National Museum of Natural Sciences, National Museums of Canada, Ottawa, are also given. Locality numbers preceded by the letter P- are the formal designations used in the National Museum catalogue of fossil vertebrates. These localities are, however, referred to by "site" numbers in the text.

History of Collecting at Scabby Butte

Fragmentary dinosaurian bones were first found at Scabby Butte as long ago as 1881 by R.G. McConnell, then field assistant to the renowned Canadian geological pioneer George M. Dawson. There is some confusion about the exact date of McConnell's original discovery. Both Selwyn (1883) and Lambe (1899) give it as 1882. However, in a report issued in May of 1882 (before that summer's field season) and covering field investigations by Dawson and McConnell in 1881, language identical to that of Selwyn states that a fossiliferous zone at "Scabby Butte ... includes large numbers of bones of vertebrates, probably dinosaurian" (Dawson, 1882, p. 5).

In any event, McConnell's specimens were unspectacular, and their occurrence having been recorded, Scabby Butte was practically forgotten in ensuing years while collectors were drawn northward to the prolific dinosurian fields along the Red Deer River (L.S. Russell, 1966). Discovery of cranial fragments of *Pachyrhinosaurus* in 1945 rekindled interest in the locality and provided the impetus for the work of the National Museum begun in 1955.

Published references to vertebrate fossils at Scabby Butte are few. Besides passing mentions in early *Reports of Progress* of the Geological Survey of Canada (Dawson, 1882, 1883, 1885), they are discussed briefly by Weston (1889) in a semi-popular account of early Survey explorations, and Lambe (1899) refers to a femur of a hadrosaur collected by McConnell. More recently, the presence of *Pachyrhinosaurus* was noted by Sternberg (1950) and L.S. Russell (1962), and Russell and Sloan (1974) have dealt with the fossil mammalian material from Scabby Butte.

Geological Setting

Scabby Butte is a small patch of badland escarpments that face north and west, about four miles northeast of the settlement of Nobleford, Alberta (Fig. 1). The locality lies at Reference UL 5732, Canada National Topographic Series, Map 82H (Edit. 1), ASE Series A 502 (Lethbridge). It appears in the southwest quarter of Canada Department of Mines and Technical Surveys airphoto A 15129-65 (Fig. 2). There erosion has dissected a low anticlinal fold in the St. Mary River Formation (see Irish, 1971). Exposures having a maximum relief of about 100 feet (30 m) occur in a restricted area, mainly in Sections 18 and 19, Township 11, Range 22 West of the fourth Meridian.

The exposed bedrock at Scabby Butte comprises mainly gray to brown clay and siltstones interbedded with lighter-coloured fine sandstones. Much of the section is carbonaceous and, with a few persistent coal-seams in the lower part, resembles rocks of the Edmonton Formation. Thus Tozer (1956) terms the beds "the Edmonton facies" of the St. Mary River Formation. At Scabby Butte, however, the coaly beds are not so sharply defined and are relatively less important in the section as a whole. There is less bentonitic material than usual in typical Edmonton sediments (= Horseshoe Canyon Formation of Irish; for redefinition of the classic Edmonton beds of authors, see Irish, 1970). Hard brown clay-ironstone nodules that are a striking feature of some Edmonton sections are not so conspicuous at Scabby Butte, although laminar ironstones occur in places. Unlike the more familiar Edmonton beds several horizons in the St. Mary River Formation throughout the general Scabby Butte area contain massive botryoidal, yellow-weathering, carbonate concretions. Broad, finely laminated, tabular cross-bedding is developed locally, mainly in light gray to white silts and sandstones in the lower beds; a few elongate sand bodies are present higher in the section. Coarse sand and conglomerate comprise only a small part of the total section. Sedimentary structures are virtually non-existent in the finer-grained material, but occasional animal borings are preserved in the sandstones.

Depositional Environment

The Cretaceous sediments at Scabby Butte were deposited during the last general retreat of the Bearpaw Sea from the Western Canadian Sedimentary Basin (McCrossan *et al.*, 1964). The situation is best described by Jeletzky (1971, p. 72): "... as the late upper Campanian Bearpaw Sea retreated the blanket of early Maestrichtian nonmarine rocks of the St. Mary River and Edmonton (including upper Brazeau) Formation spread eastward across the Southern Great Plains. Farther northeast the great lobe of the Edmonton Delta was spreading steadily southeastward, eastward, and north-eastward."

The lowest exposed rocks appear in a few gullies at the lower prairie level at the base of the escarpment. They are dark gray joint clays and shales that contain the brackish-water clam *Corbula perangulata* Whiteaves and are assigned to the basal member of the St. Mary River Formation (Tozer, 1956).

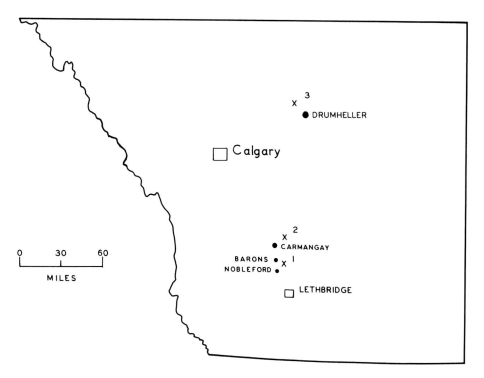

Fig. 1 Map of southern Alberta showing location of Scabby Butte (1) and other localities where *Pachyrhinosaurus* remains have been found: (2) on Little Bow River; (3) below Munson on the Red Deer River. Scabby Butte is in the St. Mary River Formation; Little Bow and Red Deer River localities are in the approximately equivalent Maestrichtian lower Edmonton (= Horseshoe Canyon) beds.

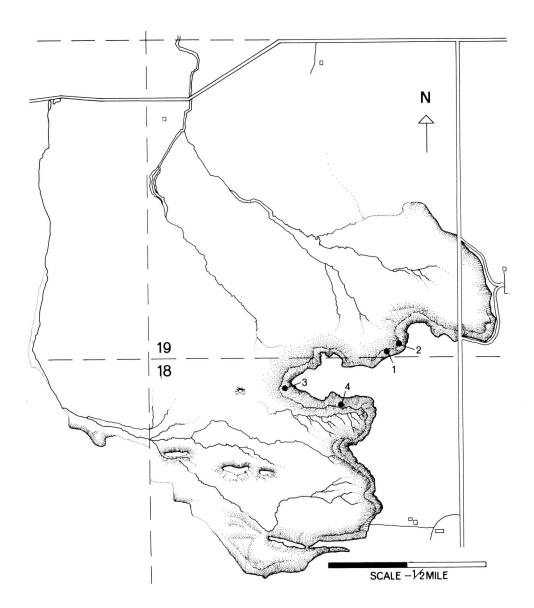

Fig. 2 Scabby Butte, showing location of Sites 1–4 of this report. Area covered is primarily section 19 and north two-thirds of Sec. 18, Tp. 11, Rge. 22, west of the 4th Meridian. Drawn from airphoto (Canada Dept. Mines and Tech. Surv., No. A 15129–65).

The lithology at and around Scabby Butte bespeaks conditions of deltaic progradation with increasing freshening of the aquatic environment through time. The linear sand bodies in the upper part of the section probably represent distributory channels on a deltaic plain; the intervening clays and silts which comprise the bulk of the strata were deposited in the interdistributory flood-basin. Carbonaceous deposits which diminish in importance towards the top of the section give evidence of improved drainage with advance of the fluvial system. The available evidence indicates that the sites of vertebrate accumulation on the whole represent non-marine depositional conditions. Localized thin tabular iron-stone beds (clay drapes?) suggest periodic flooding of subaerial surfaces, whereas massive carbonate concretions, together with marine and brackish-water fossils, attest to the persistence of nearby marine conditions.

Localities

Fossil vertebrates have been recovered from the following sites at Scabby Butte, all of which are situated in Township 11, Range 22 west of the Fourth Meridian (Fig. 2).

Site 1 (Locality P-4611)
About 100 feet (30 m) north of the south line, SE¼, SE¼, Sec. 19. This is the site from which the first cranial material of *Pachyrhinosaurus* was obtained in 1945 and 1946 (Sternberg, 1950).

Site 2 (Locality P-5510)
On the westerly escarpment in SE¼, SE¼, Sec. 19. This is the main quarry excavated in 1955 and 1957.

Site 3 (Locality P-5702)
On an erosional bench on a westward promontory near the northeast corner of NE¼, NW¼, Sec. 18. This is a "shell bed" about half way up the escarpment. It has provided most of the small vertebrate remains from Scabby Butte, including all the mammalian teeth and the vertebra of a mosasaur. Quantities of fragmentary mollusks and a few ostracods are present.

Site 4 (Locality P-5701)
South-facing escarpment near south line of NW¼, NE¼, Sec. 18. A mosasaur skeleton was found here. The locality lies at about the same stratigraphic level as the shell bed at Site 3 and about 200 yards (185 m) ESE of it. Dinoflagellates occurred sparingly near the skeleton.

Site 5 (Locality P-5706)
Northwest escarpment in NW¼, Sec. 18, about 20 feet (6 m) below upper prairie level. This locality is stratigraphically higher than the others. It provided a few bones of a large hadrosaur which, besides the mosasaur from Site 4, is the only positively associated material of one individual yet found in the Scabby Butte area.

Site 6 (Locality P-5801)
This locality includes the Scabby Butte badlands as a whole. The number is applied to all surface finds. Specimens so designated have been transported unknown distances from their sources.

The exact localities from which McConnell obtained his specimens are not recorded, but it is doubtful if any of this material occurred *in situ.*

Except at Sites 2, 3, and 4, no material of importance was found in place at Scabby Butte. Broken pieces of bones are encountered upon erosional benches and in valleys throughout the badlands, but are rarely seen on the slopes. Evidently derived from only a few horizons, such fragments rarely remain close to their source for long, making fossil prospecting at Scabby Butte a disappointing if not an exasperating experience. Bones with their long axes dipping into the banks are often broken into transverse sections that are displaced *en echelon* to the surface; also the gaps between the sections become wider in the same direction. However, bones not yet affected by weathering usually lie in the horizontal plane and segments are not separated by gaps. Evidently, repeated wetting and drying of the bentonitic matrix raises even large and heavy pieces of fossil bone to successively higher levels in the bank until they reach the surface. Once there, however, the fragments roll or slide down the slopes, often moving rapidly for considerable distances over the wet and slippery bentonitic clays. The removal of the weathered "popcorn" crust over large areas adjacent to well-preserved surface fragments usually fails to reveal their source.

Taphonomy

Among the interesting aspects of bone-bed deposits, such as those at Scabby Butte, are their differing modes of origin. The literature on the taphonomy of complete buried individuals is increasing (e.g., Gradzinski, 1969; Voorhies, 1969; Dodson, 1971; and Schäfer, 1972); but little has been done on random accumulations of disarticulated bones, representing many and varied individuals, such as occur at Sites 2 and 3. Nor is it possible to deal with these problems in much detail here, but tentative interpretations of some data obtained during work at Scabby Butte may be useful.

All vertebrate specimens at Scabby Butte have been derived from shales, mudstones, and siltstones. Bones at Site 2 lie upon a soft white sandstone, characterized by thin tabular cross-bedding, which is truncated at the interface with the overlying bone-bearing claystone. That this surface was already firm, if not consolidated, when the bones were deposited is evident from the failure of even large, heavy, and angular elements to penetrate it; neither scouring nor drifting of sand occurs around the bones. The nature of the broken edges of limb bones and the presence of "green twig" fractures in several ribs show that the bones were fresh when broken and were not reworked as fossils. Neither how nor where the animals died is, of course, known, but the close association of many elements, possibly belonging to only a few individuals suggests limited transportation. There is little evidence of preferential orientation of the bones in the quarry (Fig. 3), although some long bones tend to lie in a generally NW-SE direction. Smaller

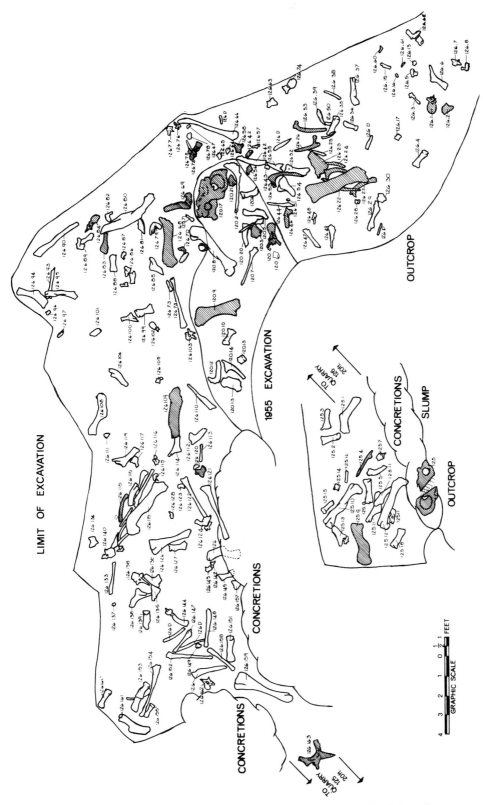

Fig. 3 Quarry diagram of Scabby Butte Site 2 (NMC locality P-5510). Shaded specimens are ceratopsian bones; all others are hadrosaurian. Numbers are field designations.

elements seem to have accumulated in the vicinity of larger bones, but in general the distribution of the specimens in the quarry appears random. The frequency of occurrence of various elements is anomalous: the limb bones recovered represent at least eight dinosaurs, but a total of only 40 vertebrae and less than 30 ribs were encountered during excavation. Except for ribs and vertebrae, the most frequently found elements were hadrosaurian pubes.

The preponderance of limb elements as compared with the originally more numerous vertebrae and ribs suggests (1) transportation of more massive bones into the burial site, (2) transportation of the vertebrae and ribs away from it, or (3) a combination of these. Bearing on this question is the tendency observed today in the subaerial disintegration of large carcasses for axial and appendicular skeletons to become separated from each other but, held together in segments by dried ligaments, to remain for some time thereafter more or less articulated (Weigelt, 1927; Toots, 1965; Müller, 1951). Sequences of articulated vertebrae, because of their projecting processes, may respond differently to both organic and physical agents of transportation than do cylindrical or flat, plate-like elements; five articulated segments comprising 100 vertebrae might, under some conditions, behave as five elements instead of 100 bones. It would not be surprising, therefore, to find an abnormally high ratio of limb bones to vertebrae in a bone-bed from which the vertebrae may have been selectively removed as intact units or, in which because of their projecting processes, they may have become anchored and remained behind while limb bones were separately deposited. The large number of hadrosaurian metapodials (21) and of fibulae suggest that limbs and feet remained partly articulated during transportation to Site 2, but later fell apart and became scattered before burial.

Experiments by Voorhies (1969) reveal a pattern of differential dispersal of different sets of elements under varying conditions of stream flow. With mammalian skeletons ribs are most often removed by saltation or flotation before limb and girdle elements, while skulls tended to remain to the last. Except for those of ceratopsians and nodosaurs, however, relatively fragile dinosaurian skulls should not be expected to undergo transportation in the mammalian fashion, and one may wonder how closely huge dinosaur bones buried in fine-grained sediments may have followed the pattern described for mammals. However, the frequency of elements at Site 2 does not appear inconsistent with Voorhies' model. It does not seem likely that the bones had been transported far, and the presence of *Pachyrhinosaurus* skulls and limb-bones suggests that the animals died nearby.

On the other hand, there is much evidence of subaerial disintegration before final burial took place. Thus fragments broken from a *Pachyrhinosaurus* skull (WL 120.17 = 9485) lie nearby, and the longitudinally split segments of a large hadrosaur femur (WL 126.80 = 9793) were scattered over an area of several square metres. The ends of limb bones are often damaged or destroyed, the relatively cancellous parts of these bones having been less resistant to weathering before burial than the dense shafts (which also better resisted deformation after burial).

A number of sets of elements (e.g., hadrosaurian pubes, dentaries, and scapulae) shows a remarkable tendency to deteriorate in particular patterns. Dentaries, regardless of size, invariably consist only of the coronoid process and about half of the horizontal ramus, and have always lost their teeth. Pubes and scapulae are represented by the thicker, proximal ends, and thinner edges are generally incomplete. Similar weathering can be observed today, for example, in artiodactyl scapulae where disintegration commences at the relatively spongy supra-scapular margin and progresses inwardly and often along the anterior and posterior borders more rapidly than on the denser surfaces. Such bones are well advanced in the fifth stage of the normal course of disarticulation as outlined by Toots (1965).

The degree to which biogenic factors may have been involved in the disintegration of bones at Site 2 is impossible to estimate. Absence of vertebrate remains, other than those of dinosaurs, in this bone-bed indicates that the site of deposition probably does not represent a biocenosis. None of the specimens shows indications of scavenging activity or of boring or cutting. However, the discovery of a few carnosaur teeth in the quarry indicates that other vertebrates may have played a part in the sorting process.

The evidence suggests that remains of hadrosaurs and ceratopsians from nearby habitats accumulated at Site 2 on a dry flood plain or long-abandoned channel floor where, unburied for some time, they underwent much subaerial weathering. The accumulation was then covered by a blanket of mud resulting from the flooding of a nearby river.

An unusual diagenetic phenomenon is the exceptional lustre displayed by most of the bones recovered at Site 2. Virtually all superficial detail has been removed from the tops and sides, and the dense, dark brown bones will occasionally reflect an image when wet! Significantly, however, the lower surfaces of the bones lying on the siltstone substrate are little polished and retain a high degree of textural detail. Broken edges are rounded off above but remain sharp below. Some fragments of ceratopsian skulls that are known to have been rugose have had all traces of sculpture polished away, producing a striking contrast when they are reunited with other pieces which were preserved with their rough surfaces facing downward. Broad, flat, upward-facing surfaces may be less polished near the centres than closer to the edges. The bones at Site 2 are overlain by about 17 feet (5 m) of joint clay. This fine-grained matrix is waxy when fresh and contains many slickensides, especially where it came in contact with resistant objects, such as concretions and bones. All stratification has been destroyed in the clays, and it seems clear that much compaction has occurred. Movement of the clay over and around the bones is thus a likely cause of their unusual polish; lower surfaces of the bones resting close to the relatively immobile substrate were protected from abrasion by the overlying bony mass. It is interesting that bones enclosed in massive botryoidal concretions at Site 2 are less polished than those imbedded in claystone, indicating that most compaction occurred in the clays after the beginning of carbonate concentration in the nodular zones. Bones surrounded by thin-layered concretionary carbonate show the characteristics of bones in the clays, and it is obvious that this material is of more recent origin than the massive concretions. That they formed after compaction had ceased is shown by their roughened outer surfaces; and it is probable that this episode of precipitation is still in progress.

Scabby Butte Site 3 is a "shell bed" of a kind commonly encountered in the late Cretaceous and early Tertiary deposits of western Canada. Such accumulations are much sought as sources of remains of small mammals which seem to occur in them with more than usual frequency. Most often, shell beds contain, in addition to indifferently preserved molluscan shells, a variety of small bones and teeth. That Site 3 represents a winnowed deposit seems probable, and considerable transportation of some of the material is evinced by well-rounded fragments of dinosaur bone and teeth and by a mosasaur vertebra which, in all probability, was derived from the skeleton collected at Site 4. The associated molluscs, though abundant, are too badly broken for positive identification, but the fresh-water unionid *Pleisielliptio* is probably present, and two fresh-water ostracods *Cypridea (Cypridea)* sp. and *Candona* sp. noted above occur sparingly in the deposit. By far the most common vertebrate remains are teeth of the fresh-water ray *Myledaphus*. Not improbably, their occurrence in the shell bed reflects an association with the molluscs in life, and the presence of phyllodont crushing teeth in the deposit may be similarly explained. Although not widely accepted, the idea that the remains of small vertebrates, including mammalian teeth, in deposits of this sort were derived ultimately from coprolites seems to me a reasonable possibility; one may suppose that small mammals might have fallen prey to lurking champsosaurs or fish, while investigating clams exposed in shallow water. Crocodilians, however, were probably not involved as virtually all ingested osseous material would likely have been destroyed in their digestive tracts.

The discovery of a mosasaur skeleton at Site 4 deserves special mention as possibly the first report of these supposedly strictly marine animals in non-marine sediments. A short segment of the trunk, consisting of three vertebrae, about 18 costal cartilages and both articulated forelimbs and flippers, lay beneath an articulated sequence of caudal vertebrae. These conditions are suggestive of desiccation and partial dismemberment before burial (Schäfer, 1972; Toots, 1965). Although it is impossible to know how much of the skeleton was originally present, weathered vertebrae and other scrap in its vicinity indicate that much had been destroyed before recovery. The vertebra recovered at Site 3 is of appropriate size and, since Sites 3 and 4 appear to be at about the same stratigraphic level, this bone may have been derived from the skeleton at Site 4 *before* burial. The bones were partly imbedded in a thin (¼ ", 7 mm) coaly substrate and covered by a waxy, brittle, brown claystone exhibiting massive jointing. A thin sand lens occurred just above the skeleton. The costal cartilages are preserved in approximate articulation, as is common in mosasaurs.

The presence of a mosasaur in presumably non-marine strata raises questions for which plausible explanations are both speculative and numerous. Is it possible that the animal was accustomed to fresh-water habitats? L.S. Russell (1931) has suggested this possibility for plesiosaurs, but he was able to cite several suggestive occurrences, whereas I know of no other record of mosasaur bones from non-marine deposits. Deltaic depositional environments are so varied that any number of hypothetical models could be suggested to explain the available taphonomic and sedimentary evidence. The assumption that marine organisms somehow became buried in non-marine sediments is consistent with current ideas about estuarine and lagoonal environments (Emery and Stevenson, 1957). The obvious suggestion that the mosasaur-bearing sediments at this spot were in fact not fresh water, but marine in origin, is supported by the discovery of dinoflagellates near the skeleton. A fluctuating saline wedge might have allowed the mosasaur and the dinoflagellates to reach an essentially fresh-water environment sufficiently distant from their customary habitat to have prevented their return.

The recent analysis of the depositional environments of the lower Edmonton beds in the Drumheller badlands by Shepheard and Hills (1970) indicates that the marine influence was much more significant in the development of Edmonton-type lithologies than has been generally supposed. The St. Mary River deposits developed at Scabby Butte and vicinity may conform in a general way to their "lower floodplain" model with some characteristics of lagoonal and estuarine environments. Some such setting would, presumably, have permitted the entomb-ment of the mosasaur at Site 4, possibly as a result of a single brief episode of flooding by marine waters which left no other evidence of the event except the mosasaur and the dinoflagellates.

In summary, taphonomic and depositional evidence at Scabby Butte contrasts with that described by Dodson (1971) for the Oldman Formation on the Red Deer River, Alberta. There more or less complete and individual dinosaur burials were "common in channel sediments and rare in overbank deposits," whereas the reverse seems to hold at Scabby Butte. The Site 2 accumulation, with a pre-ponderance of only two taxa, suggests local catastrophic rather than "regular" events as inferred from the "persistence of the characteristic sedimentary associa-tion of fossils over a 200 ft. stratigraphic interval" (Dodson, 1971, p. 21).

The Scabby Butte Fauna

At Scabby Butte 32 vertebrate taxa, including 26 identified to the generic level, have been found (Table 1). A comparison of this assemblage with the five best-known non-marine late Cretaceous faunas from western North America (Table 2) reveals fairly close correspondence between the Scabby Butte and the other assemblages. However, Scabby Butte, lacking the great variety of dinosaurs of the Oldman and lower Edmonton, and most of the amphibian and saurian elements of the Upper Milk River, Lance, and Bug Creek faunas, remains the least diverse of the six assemblages. Since sampling for small vertebrates at Scabby Butte has been confined to a few hundred pounds of matrix from one locality, taxonomic variety, particularly among the mammals, seems remarkable when compared with the much more broadly sampled Milk River and Lance strata. For an example of the effect of screening on the known fauna of the Canadian dinosaur beds, compare L.S. Russell's 1964 summary of the Upper Milk River fauna with 13 taxa and the summary by Fox in 1972 listing 63, an increase of 50, largely additional micro-vertebrates. Comparable increases in the Oldman and earlier Edmonton assem-blages can be confidently anticipated when washing techniques for small specimens are widely applied to these deposits.

Although one is tempted to analyse the Scabby Butte assemblage along the palaeoecological lines suggested by Estes and Berberian (1970) for the Bug Creek fauna, such an exercise, based on the limited sample presently available,

Table 1. The Scabby Butte vertebrates. (Identifications for class Mammalia by R.E. Sloan [Multituberculata] and L.S. Russell [Theria] 1974).

Class **ELASMOBRANCHII**
Order Isuriformes
  Orectolobidae
    Squatirhina americana Estes
Order Squatinoidei
  Squatinidae
    Squatina Risso, sp. indet.
Order Rajiformes
  Dasyatidae
    Myledaphus bipartitus Cope

Class **OSTEICHTHYES**
Order Amiiformes
  Amiidae
    Amia cf. *fragosa* (Jordan)
Order Aspidorhynichiformes
  Aspidorhynchidae
    Belonostomus cf. *longirostris* (Lambe)
Order Lepisosteiformes
  Lepisosteidae
    Lepisosteus Lacépède, sp. indet.
Order Elopiformes
  Phyllodontidae
    Paralbula casei Estes
Order Perciformes
  Sciaenidae
    Platacodon cf. *nanus* Marsh

Class **AMPHIBIA**
Order Urodela
  Batrachosauroididae
    Opisthotriton Auffenberg, sp. indet.

Class **REPTILIA**
Order Testudines
  Trionychidae
    Aspideretes Hay, sp. indet.
  Baenidae
    Boremys Lambe, sp. indet.
Order Eosuchia
  Champsosauridae
    Champsosaurus Cope, sp. indet.

Order Squamata
  Mosasauridae
    Plioplatecarpus Dollo, sp. indet.
Order Crocodilia
  Crocodylidae
    Leidyosuchus Lambe, sp. indet.
    Crocodilia, gen. et sp. indet.
Order Saurischia
    Tyrannosauridae, gen. et sp. indet.
    ?Coeluridae, gen. et sp. indet.
    Ornithomimidae, gen. et sp. indet.
  Troodontidae
    Troodon Leidy, sp. indet.
Order Ornithischia
  Hadrosauridae
    ?*Edmontosaurus* Lambe, sp. indet.
  Ceratopsidae
    Pachyrhinosaurus canadensis Sternberg
  Nodosauridae
    Edmontonia cf. *longiceps* Sternberg

Class **MAMMALIA**
Order Multituberculata
  Cimolomyidae
    Cimolomys gracilis Marsh (*sensu* Clemens 1963)
    Meniscoessus conquistus Cope
  Ectypodontidae
    Mesodma cf. *thompsoni* Clemens
Order Marsupialia
  Didelphidae
    Pediomys cf. *cooki* Clemens
    Pediomys cf. *krejcii* Clemens
  Stagodontidae
    Didelphodon? sp.
    Eodelphis? sp.
Order Insectivora
  Cimolestidae
    Cimolestes sp.
  Adapisoricidae
    Gypsonictops sp.
Order Carnivora
  Miacidae?, gen. et sp. indet.

would command little confidence. The evidence seems only adequate to suggest that the Scabby Butte assemblage probably represents a mixture of contemporaneous late Cretaceous fresh-water aquatic to semi-aquatic and the so-called distal, mainly dinosaurian, communities. In this, the Upper Milk River and Bug Creek faunas are not dissimilar. The Bug Creek interpretation (Estes and Berberian, 1970), with its restriction of terrestrial and riparian habitats, seems to fit the Scabby Butte situation better than the "wooded swamp habitat, with small to medium-sized watercourses and some ponding," which these authors postulate for their Lance locality v-5620.

Table 2. Late Cretaceous vertebrate faunas of western North America compared with the Scabby Butte assemblage. Sources: Langston, 1965; L.S. Russell, 1964; Estes and Berberian, 1970; Sloan and Van Valen, 1965; Fox, 1972; Sloan and Russell, 1973; L.S. Russell, *pers. comm.*, 1973; Lillegraven, 1969; Clemens, 1973.

Scabby Butte	Upper Milk River Formation	Oldman Formation	Lower and Middle Edmonton (=Horseshoe Canyon) Formation	Lance Formation (incl. =U. Edmonton; Frenchman)	Hell Creek (Bug Creek Fauna)
Squatirhina	—	—	—	X	—
Squatina	—	—	—	X	—
Myledaphus	X	X	X	X	X
Amia	X	—	—	X	X
Belonostomus	X	X	X	—	X
Lepisosteus	X	X	X	X	X
Paralbula	X	X	—	—	X
Platacodon	X	—	—	X	X
Opisthotriton	X	—	—	X	X
Champsosaurus	X	X	X	X	X
Aspideretes	X	X	X	X	—
Boremys	—	X	—	—	—
Leidyosuchus	X	X	X	X	X
Tyrannosauridae	X	X	X	X	X
Coeluridae	X	X	X	X	X
Ornithomimidae	X	X	X	X	—
Troodon	—	X	X	—	—
Edmontosaurus	—	—	X	—	—
Pachyrhinosaurus	—	—	X	—	—
Anchiceratops	—	X	X	—	—
Edmontonia	—	X	X	—	—
Cimolomys	X	X	—	X	X
Miniscoessus	X	—	—	X	X
Mesodma	X	X	—	X	X
Pediomys	X	—	—	X	X
Didelphodon(?)	—	—	—	X	X
Eodelphis(?)	X	X	—	X	—
Cimolestes	—	—	—	X	X
Gypsonictops	—	—	—	X	X

Palaeoecological Interpretations

The vertebrate fauna from Scabby Butte was predominantly fresh-water to terrestrial in habits, but, as noted, the presence of marine organisms (shark, mosasaur, dinoflagellates) gives evidence of saline or brackish conditions in the vicinity. Most interesting are the ecological implications of pollen and spores obtained from matrix removed from the interior of a *Pachyrhinosaurus* skull (WL 120.17 = 9485) at Site 2. This sample contains the following 28 taxa, identified by S.K. Srivastava:

Balmeisporites bellus Kondinskaya
B. kondinskayae Srivastava and Binde
Cicatricosisporites sp.
Hamulatisporis amplus Stanley
H. hamulatis Krutzsch
Lycopodiumsporites sp.
Polypodiidites sp.
Ginkgocycadephytus sp.
Picea sp.
Glyptostrobus sp.
Aquilapollenites ?macgregorii Srivastava
A. spinulosus Funkhouser
A. dolium (Samoilovitch) Srivastava
A. decorus Srivastava
A. validus Srivastava

A. polaris Funkhouser
A. ?aptus Srivastava
A. augustus Srivastava
A. drumhellerensis Srivastava
A. hirsutus Srivastava
Mancicorpus albertensis Srivastava
Mancicorpus sp.
Trifossapollenites ellipticus Rouse
Erdtmanipollis procumbentiformis (Samoilovitch) Srivastava
Cranwellia rumseyensis Srivastava
Momipites sp.
Coriaripites sp.
Liliacidites sp.

Momipites, *Coriaripites*, and *Liliacidites* are rare; *Glyptostrobus* pollen occurs in fair abundance, but other coniferous pollen is rare. Only one grain of *Trifossapollenites ellipticus* Rouse was located in the sample. All other taxa are well represented.

Of this assemblage, Srivastava writes (pers. comm., 1973):

Ecological implications: The abundance of pollen of *Glyptostrobus* indicates that such plants were forming large open stands on marshes. There are also pollen related to those of *Taxodium* (Swamp Cypress) which today forms open clusters on swampy areas in the lower Mississippi valley and the Gulf Coast. *Glyptostrobus* and *Taxodium* are both conifers (family – Taxodiaceae). *Cranwellia* has a close affinity with the pollen of the extant genus *Elytranthe* – an epiphyte presently distributed in southeastern Asia and western Malaysia. The plants producing *Cranwellia* might have been epiphytic on *Glyptostrobus* or *Taxodium*, and may have been a chief source of foliage for herbivorous animals of that time. *Aquilapollenites* – pollen of some unknown angiosperms, may also be related to the pollen of the epiphytic family Loranthaceae. The affinities of *Cranwellia* and *Aquilapollenites* with the modern genus *Elytranthe* and the family Loranthaceae respectively are based on the morphological similarities of fossil and modern pollen. The epiphytic relation of plants producing these fossil pollen with conifer trees is conjectural and only based on the association of these pollen in assemblages and present day observations. In summary, the immediate area ... was most likely an open subtropical forest, with the trees supporting epiphytes, and the surface swampy and marshy, containing some ferns. A very few pollen grains of either Cycads or *Ginkgo* are present, suggesting that the number of plants of the cycadophytes (Bennetitales and/or Cycadales), or *Ginkgo* was small in this immediate locale at this particular time.

Most of the botanical evidence seems to favour humid subtropical conditions like those inferred for the lower part of the Edmonton sequence (Srivastava, 1972).

Age and Correlation of Scabby Butte Vertebrate-Bearing Beds

Jeletzky (1971) attributes an early Maestrichtian age to the non-marine St. Mary River and Edmonton (including the upper Brazeau) Formations. L.S. Russell (1964) equates the St. Mary River with his Edmontonian stage, which he believes is equivalent to the lower part of the Maestrichtian. Others (e.g. Gill and Cobban, 1966; Tozer, 1956) would assign at least part of the St. Mary River beds a slightly older, latest Campanian, age. Tozer's (1952) recognition of a Kneehills Tuff equivalent at the top of the St. Mary River section on the Oldman River shows that here at least the upper limit of the formation occurs lower than the non-marine Lancian Stage as defined by L.S. Russell (1964). Southwest of Edmonton the age of the Kneehills Tuff is about 65MY (Follinsbee *et al.*, 1961) and thus falls within Maestrichtian time (Gill and Cobban, 1973). Nothing resembling the Kneehills Tuff or Whitemud beds has been observed at Scabby Butte, and the presumption is that Scabby Butte deposits lie lower in the section. There is no doubt that the St. Mary River Formation is homotaxial from west to east and, in the absence of radiometrically datable rock at Scabby Butte, age assignment must rest upon its fossil content. Ammonites are absent from the local section, and the other molluscs (the bivalves *Corbula perangulata*, *Unio stantoni*, cf. *Plesielliptio*, and a gastropod *Viviparus* sp.) are long-ranging taxa of little help in correlation within the late Campanian/early Maestrichtian interval. Two ostracods, *Cypridea* (*Cypridea*) sp. and *Candona* spp., both found at Site 3, range from the Middle Jurassic to the Maestrichtian (Levinson, pers. comm., 1973).

At present, the strongest direct indication of the age of rocks intimately associated with the Scabby Butte vertebrates are the pollen of *Aquilapollenites* and *Mancicorpus*. These are believed to be especially significant for intercontinental correlation of Maestrichtian deposits (Srivastava, 1972).

There is general agreement that the St. Mary River Formation as a whole is roughly synchronous with the Edmonton Group as defined by Irish (1971). Stratigraphic and facies relationships are described by L.S. Russell (1950): 15 miles (24 km) north of Scabby Butte, near Carmangay, distinctive Edmonton and St. Mary River lithologies interfinger along the Little Bow River (L.S. Russell, 1932; Irish, 1971). Tozer (1956, p. 16), reviewing the stratigraphic and molluscan evidence, concludes that "... correlation of the lower Edmonton with the St. Mary River ... is in harmony with the available evidence. ..."

Although the fossil megafloral assemblage has not been analysed, the palynological sample from Site 2 indicates a probable correlation with the "upper part of the lower Edmonton and the lower part of the middle Edmonton below the Mauve-shale containing the Kneehills Tuff" (Srivastava, pers. comm., 1969). Although correlation based upon first appearance of taxa in the Edmonton section shows the best correspondence with Srivastava's assemblage Zones I to III (with, however, one "species-restricted" representative of Zone v), all taxa except *Mancicorpus albertensis* range upward to near the top of Zone iv or higher; the strongest implied palynomorph correlation is with Srivastava's assemblage Zones iv and v, which fall within Ower's subdivisions a-c of the Edmonton Formation, or about the upper two-thirds of the Horseshoe Canyon Formation (Srivastava, 1970; Ower, 1960; Irish, 1970). The first and last appearances of Scabby Butte microfloral taxa in the Edmonton section are: Assemblage Zone i (transition zone)—*Aquilapollenites ?aptus*[VI], *A. drumhellerensis*[IV], *A. polaris*[IX]; Zone ii (*Aquilapollenites leucocephalus* zone)—*A. augustus*[VIII], *A. decorus*[IV], *A. ?macgregorii*[VI], *Cranwellia rumseyensis*[VI], *Manicorpus albertensis*[IV], *Trifossapollenites ellipticus*[IV]; Zone iii (*Wodehousea jacutense* and *W. gracile* zone)—*Aquilapollenites dolium*[VI], *A. hirsutus*[VI], *A. spinulosus*[V], *A. validus*[V]; Zone iv (*Mancicorpus vancampoi* zone)—*Erdtmanipollis procumbentiformis*[V] (species restricted).

Perhaps the most restrictive non-marine stratigraphic indicator presently available from these levels is the dinosaur *Pachyrhinosaurus*: the sole known species *P. canadensis*, abundant at Scabby Butte and in the exposures on Little Bow River, also occurs only in beds of Ower's Edmonton a in the Drumheller district (D.A. Russell and Chamney, 1967). *Pachyrhinosaurus* was presumably a rapidly evolved genus, like other ceratopsians, whose distribution in time could hardly involve measurable homotaxis (Langston, 1967). Its occurrence at various places in the western interior of Canada, therefore, may be accepted in the absence of contrary evidence as a key to refined correlation in this region. Thus the available evidence suggests that the Scabby Butte deposits are correlative with the middle and upper parts of Ower's Edmonton a, and are of early Maestrichtian age.

Acknowledgments

I wish to acknowledge with special thanks the help of the late Mr. Harold Shearman, at the National Museum of Canada, for his assistance in the field and for his expert preparation of most of the Scabby Butte material collected by my field parties. His assistance rendered during several visits to Ottawa for study of the collection has been most helpful. Aid and counsel from the following gentlemen is also gratefully acknowledged: Dr. E.T. Tozer of the Geological Survey of Canada for his identification of molluscs, Dr. S.K. Srivastava of the Chevron Oil Field Research Company for extensive data on the palynomorphs, Dr. Stuart D. Levinson of Esso Production Research Corporation for his identifications and comments on the ostracods, and Dr. Richard Estes of San Diego State University for his assistance with certain of the fragmentary microvertebrates. I have profited from discussions with Drs. D.A. Russell and C.M. Sternberg. Illustrations are by Mrs. Doris Tischler.

Literature Cited

CLEMENS, W.A.

1973   Fossil mammals of the type Lance Formation Wyoming, III. Eutheria and summary. Univ. Calif. Pubs. Geol. Sci., 94: 1–102.

DAWSON, G.M.

1882   Preliminary note on the geology of the Bow and Belly River districts, N.W. Territory, with special reference to coal deposits. Geol. Surv. Canada, Dawson Bros., Montreal (1882): 1–19.

1883   Preliminary report on the geology of the Bow and Belly River region, N.W. Territory, with special reference to the coal deposits. Rept. of Progress for 1880–81–82, Geol. and Nat. Hist. Surv. Canada, part B: 1–23.

1885   Report on the region in the vicinity of the Bow and Belly Rivers, North-West Territory. Rept. of Progress for 1882–83–84, Geol. Surv. Canada, part C: 1–169.

DODSON, P.

1971   Sedimentology and taphonomy of the Oldman Formation (Campanian), Dinosaur Provincial Park, Alberta (Canada). Palaeogeogr., Palaeoclimatol., Palaeoecol., 10: 21–74.

EMERY, K.O. AND R.E. STEVENSON

1957   Estuaries and lagoons. Treatise on marine ecology and paleoecology, I. Ecology, J.W. Hedgpeth, ed., Geol. Soc. Amer. Mem., 67: 673–749.

ESTES, R. AND P. BERBERIAN

1970   Paleoecology of a late Cretaceous vertebrate community from Montana. Brevoria, 343: 1–35.

FOLINSBEE, R.E., H. BAADSGAARD AND J. LIPSON

1961   Potassium-argon dates of Upper Cretaceous ash falls, Alberta, Canada. In Kulp, J.L., ed., Geochronology of rock systems, New York Acad. Sci. Annals, 91: 352–363.

FOX, R.C.

1972   A primitive therian mammal from the Upper Cretaceous of Alberta. Canadian Jour. Earth Sci., 9: 1479–1494.

GILL, J.R. AND W.A. COBBAN

1966   The Red Bird section of the Upper Cretaceous Pierre Shale in Wyoming. U.S. Geol. Surv. Prof. Pap. 393-A: A1–A73.

1973   Stratigraphy and geologic history of the Montana Group and equivalent rocks, Montana, Wyoming, and North and South Dakota. U.S. Geol. Surv. Prof. Pap. 776: 1–37.

GRADZINSKI, R.

1970   Sedimentation of dinosaur-bearing Upper Cretaceous deposits of the Nemegt Basin, Gobi Desert. Palaeontologia Polonica, 21: 147–229.

IRISH, E.J.W.

1970   The Edmonton Group of south-central Alberta. Bull. Canadian Petrol. Geol., 18: 125–155.

1971   Southern plains of Alberta, west of fourth meridian: Map 1286A. Ottawa: Geol. Surv. Canada.

JELETZKY, J.A.

1971   Marine Cretaceous biotic provinces and paleogeography of western and arctic Canada: illustrated by a detailed study of ammonites. Geol. Surv. Canada Pap. 70–22: 1–92.

LAMBE, L.M.
  1899   On reptilian remains from the Cretaceous of North-western Canada. Ottawa
          Naturalist, 13: 68–70.

LANGSTON, W., JR.
  1965   Pre-Cenozoic vertebrate paleontology in Alberta: its past and future. Vert. Pal.
          in Alberta, Univ. Alberta, 1965: 9–13.
  1967   The thick-headed ceratopsian dinosaur *Pachyrhinosaurus* (Reptilia: Ornithischia)
          from the Edmonton Formation near Drumheller, Canada. Canadian Jour. Earth
          Sci., 4: 171–186.
  In press   The ceratopsian dinosaurs and associated lower vertebrates from the St. Mary
          Formation (Maestrichtian) at Scabby Butte, southern Alberta. Canadian Jour.
          Earth Sci.

LILLEGRAVEN, J.A.
  1969   Latest Cretaceous mammals of upper part of Edmonton Formation of Alberta,
          Canada, and a review of marsupial-placental dichotomy in mammalian evolution.
          Univ. Kansas Paleon. Contrib., art. 50: 1–122.

MC CROSSAN, R.G. *et al.*
  1964   Geological history of western Canada. Alberta Soc. Petrol. Geol.: i–x, 1–232.

MÜLLER, A.H.
  1951   Grundlagen der Biostratonomie. Abh. Akad. Wiss. Berlin, Math-Naturwiss. Kl.
          1950: 1–147.

OWER, J.R.
  1960   The Edmonton Formation. Jour. Alberta Soc. Petrol. Geol., 8: 309–323.

RUSSELL, D.A. AND T.P. CHAMNEY
  1967   Notes on the biostratigraphy of dinosaurian and microfossil faunas in the Edmon-
          ton Formation (Cretaceous), Alberta. Nat. Hist. Pap., Nation. Mus. Canada, 35:
          1–22.

RUSSELL, L.S.
  1931   Fresh water plesiosaurs. Canadian Field Nat., 45: 135–137.
  1932   The Cretaceous-Tertiary transition of Alberta. Trans. Roy. Soc. Canada, (3)IV,
          26: 121–156.
  1950   Correlation of the Cretaceous-Tertiary transition in Saskatchewan and Alberta.
          Bull. Geol. Soc. Amer., 61: 27–42.
  1962   Mammal teeth from the St. Mary River Formation (Upper Cretaceous) at Scabby
          Butte, Alberta. Nat. Hist. Pap., Nation. Mus. Canada, 14: 1–4.
  1964   Cretaceous non-marine faunas of northwestern North America. Life Sci. Contrib.,
          Roy. Ont. Mus., 61: 1–24.
  1966   Dinosaur hunting in western Canada. Life Sci. Contrib., Roy. Ont. Mus., 70:
          1–37.

SCHÄFER, W.
  1972   Ecology and palaeoecology of marine environments: translated from the German
          Actuo-Paläontologie by Wilhelm Schäfer (1962). Univ. Chicago Press, Chicago,
          xii and 568 pp.

SELWYN, A.R.C.
  1883   Summary reports of the operations of the Geological Corps to 31st December,
          1881, and to 31st December, 1882. Rept. of Progress for 1880–81–82, Geol. and
          Nat. Hist. Surv. Canada: 1–45.

SHEPHEARD, W.W. AND L.V. HILLS
  1970   Depositional environments, Bearpaw-Horseshoe Canyon (Upper Cretaceous)
          transition zone, Drumheller "Badlands", Alberta. Bull. Canad. Petrol. Geol., 18:
          166–215.

SLOAN, R.E. AND L.S. RUSSELL
1965   Cretaceous mammals from Montana. Science, 148: 220–227.

SLOAN, R.E. AND L.S. RUSSELL
1974   Mammals from the St. Mary River Formation (Cretaceous) of southwestern Alberta. Life Sci. Contrib., Roy. Ont. Mus., 70: 1–22.

SRIVASTAVA, S.K.
1970   Pollen biostratigraphy and paleoecology of the Edmonton Formation (Maestrichtian), Alberta, Canada. Palaeogeogr., Palaeoclimatol., Palaeoecol., 7: 221–276.
1972   Paleoecology of pollen-genera *Aquilapollenites* and *Mancicorpus* in Maestrichtian deposits of North America. 24th Internation. Geol. Congress, sect. 7: 111–120.

STERNBERG, C.M.
1950   *Pachyrhinosaurus canadensis*, representing a new family of Ceratopsia, from southern Alberta. Bull. Nation. Mus. Canada, 118: 109–120.

TOOTS, H.
1965   Sequence of disarticulation in mammalian skeletons. Univ. Wyo., Contrib. Geol., 4: 37–39.

TOZER, E.T.
1952   The St. Mary River-Willow Creek contact on Oldman River, Alberta. Geol. Surv. Canada Pap. 52–53: 1–9.
1956   Uppermost Cretaceous and Paleocene non-marine molluscan faunas of western Alberta. Geol. Surv. Canada Mem. 280: 1–125.

VOORHIES, M.R.
1969   Taphonomy and population dynamics of an early Pliocene vertebrate fauna, Knox County, Nebraska. Univ. Wyo., Contrib. Geol., Spec. Pap. 1: 1–69.

WEIGELT, J.
1927   Rezente Wirbeltierleichen und ihre paläobiologische Bedeutung. Max. Weg, Leipzig, xvi and 227 pp.

WESTON, T.C.
1889   Reminiscences among the rocks. Warwick Bros. and Rutter, Toronto, ix and 328 pp.

The Ecology of Dinosaur Extinction

Robert E. Sloan
Professor, Department of Geology and Geophysics,
University of Minnesota, Minneapolis

Abstract

While the late Cretaceous extinction of dinosaurs was essentially simultaneous, at least throughout North America, it was the result of circumstances that persisted for at least 12 million years. The stratigraphic distribution of the members of the *Triceratops* community and its successors of the western interior of North America casts doubt upon hypotheses of catastrophic extinction. There are changes in abundance but no significant taxonomic or morphologic changes in the species of the *Triceratops* community in the last 1 to 2 million years (uppermost 100 feet) of its existence; hypotheses requiring rapid rates of inadaptive evolution may thus be rejected. A humid, warm temperate to sub-tropical, angiosperm-dominated forest community containing the *Triceratops* fauna was replaced by a cooler, temperate, more coniferous forest community containing the rapidly evolving *Protungulatum-Stygimys* fauna. Both of these forest communities were in existence simultaneously during the deposition of the uppermost 80 feet of the Hell Creek Formation. In view of the progressive reduction in numbers of dinosaurs and the simultaneous changes in plants over this stratigraphic interval, a change to a cooler, more continental climate provides both a sufficient and an observed cause for the extinction of this community of dinosaurs.

Of all the problems of palaeontology, one of the most interesting is that of extinctions. There is a popular fascination with the possible reasons for the total extinction of a once flourishing group of organisms. Palaeontologists are continually being asked by lay friends and students to explain the reason (or reasons) for the extinction of some animal. If one is lucky, friends will ask about some extinction for which the reasons either are known or seem intuitively obvious. However, the animals whose extinction arouses most curiosity are dinosaurs, especially their final late Cretaceous extinction. Some notion of the various hypotheses that have been proposed is given by Jepsen (1964):

Authors with varying competence have suggested that dinosaurs disappeared because the climate deteriorated (became suddenly or slowly too hot or cold or dry or wet), or that the diet did (with too much food or not enough of such substances as fern oil; from poisons in water or plants or ingested minerals; by bankruptcy of calcium or other necessary elements). Other writers have put the blame on disease, parasites, wars, anatomical or metabolic disorders (slipped vertebral discs, malfunction or imbalance of hormone and endocrine systems, dwindling brain and consequent stupidity, heat sterilization, effects of being warm-blooded in the Mesozoic world), racial old age, evolutionary drift into senescent overspecialization, changes in the pressure or composition of the atmosphere, poison gases, volcanic dust, excessive oxygen from plants, meteorites, comets, gene pool drainage by little mammalian egg-eaters, overkill capacity by predators, fluctuation of gravitational constants, development of psychotic suicidal factors, entropy, cosmic radiation, shift of Earth's rotational poles, floods, continental drift, extraction of the moon from the Pacific Basin, drainage of swamp and lake environments, sunspots, God's will, mountain building, raids by little green hunters in flying saucers, lack of even standing room in Noah's Ark, and paleoweltschmerz.

The character of the problem is thus obvious, although the reasons for extinction are not. The history of this problem has been one of many *ad hoc* hypotheses, upon which the facts of the situation have exerted little or no control.

Species become extinct by failing to reproduce as rapidly as their members die, or by a reduction in population density or an increase in intensity of selection. Extinction of species is the rule, not the exception. The likelihood that given species will have any descendants decreases more or less exponentially with time. In general, faunas or communities become extinct species by species.

Extinction is a process which almost always involves progressive range restrictions over a period of time. These range restrictions are generally the result of a partial breakdown of a community and may be induced either by major changes in the physical variables of the community or by changes within the trophic levels of the ecosystem itself. The latter may be due to any or all of (1) increase in predation by elements of a higher trophic level, (2) increase in competition by elements of the same trophic level, or (3) changes in elements of lower trophic levels. Any decision as to which of these alternatives are applicable in a single case calls for detailed observations on occurrence and associations which have not, in general, been carried out.

The importance and extent of the late Cretaceous extinctions are not as great as has generally been supposed. They are of a spectacular nature only because of the quasi-simultaneous extinction of the last surviving members of larger taxa which had been dwindling in variety for some 2 to 5 million years. These taxa are the reptilian orders and families Ornithischia, Saurischia, Pterosauria, Mosasauridae, Sauropterygia, Ichthyosauria and Protostegidae, two family-level taxa of marsupials, Pediomyinae and Stagodontidae, and various families of invertebrates as summarized by Newell (1962).

These extinctions are the result of the partial collapse of food webs caused by a reduction in the variety and amount of primary production by photosynthetic plants. This took place simultaneously in two major groups of communities—the open ocean planktonic-nektonic community, and a series of lowland coastal plain communities.

Tappan (1968) and Bramlette (1965) have addressed themselves to the problem of the marine extinctions. Tappan, further documented by Lipps (1970), suggests that marine extinctions at the end of the Cretaceous involved the collapse of a food web based on phytoplankton, and documents the reduction of phyto-plankton. This food web involves planktonic Foraminifera as major components; in the trophic level next above the phytoplankton only three to six of the eighteen late Cretaceous genera of planktonic Foraminifera survived into the early Paleocene (Danian). Among the higher trophic levels of these communities which also suffered reduction or extinction were some teleosts and all ammonites, belemnites, plesiosaurs, ichthyosaurs, and mosasaurs. Tappan (1968) follows Bramlette (1965) in suggesting that the extinction at the end of the Cretaceous of the calcareous nanoplankton resulted from nutrient depletion in shallow seas because of reduced elevations of the continents and sluggish circulation in shallow seas. This seems unlikely, however, since the sea level underwent a net lowering of some 100–200 feet (33–66 m) at the end of the Cretaceous, as indicated by the common disconformity at the Maastrichtian-Danian contact in marine sedi-ments and the widespread marine regression in the Maastrichtian in the western interior of North America. Thus inorganic nutrient flow to the ocean would have increased. The ultimate reasons for the collapse of a major portion of the higher trophic levels of the late Cretaceous planktonic-pelagic communities are not yet clear. I shall attempt to show that the extinction of at least one con-temporary dinosaur community can be explained by a reduction in and change of character of the primary productivity.

Dinosaur extinction was hardly instantaneous. While few details of distribution of the latest Cretaceous dinosaurs are known for most continents, it is known that at least in North America the process took place over about 12 million years. During this time there was a steady and progressive decline in the number of taxa of dinosaurs. The Campanian Judith River and Oldman faunas of Montana and Alberta (Sahni, 1972; Ostrom, 1961; Lull, 1933; Lull and Wright, 1942) com-prise the most varied dinosaur faunas of all time. The approximate age of these faunas is 75 million years. There was a progressive reduction in dinosaur diversity up to the very end of the Cretaceous when the Kirtland, North Horn, Lance, Hell Creek, and Upper Edmonton faunas existed. These several faunas are all closely contemporaneous but show considerable geographic diversity. While these faunas all become extinct at the end of the Cretaceous, approximately 64 million years ago (L. S. Russell, 1973), there is some evidence that the extinction occurred slightly earlier in the north than in the south.

The latest Cretaceous coastal plain communities are greatly varied even in North America, let alone the world. The community on the western slope of the Cordillera, represented by specimens from the Moreno Formation of California, is very poorly known, as are those of the eastern half of North America. We know of three contemporary communities on the coastal plain of the eastern side of the Cordillera and the western shore of the Bearpaw-Pierre-Upper Mancos Sea. They occur from (the present) 52° to 29° N latitude. As expected, there is considerable diversity of community structure as a function of latitude (Sloan, 1969). Two of these are dinosaur communities, the southern *Alamosaurus* community and the more northern *Triceratops* community. The third is the *Protungulatum-Stygimys* community, which includes the ancestors of most Tertiary mammals. The southernmost of these communities occurs in the Javalina Formation of the Big

Bend National Park of southern Texas, in the Kirtland and Fruitland formations of the San Juan Basin of New Mexico, and in the North Horn Formation of central Utah. The known latitudinal range of this community is from 29° to 39° N. This community includes the principal sauropod in the late Cretaceous of North America, *Alamosaurus sanjuanensis*; as well as a horned dinosaur, *Pentaceratops sternbergi*; a flat-headed duckbill dinosaur, *Kritosaurus navajovius*; and a crested duckbill dinosaur, *Parasaurolophus tubicen*.

The second and more northerly dinosaur community is known to occur in the Denver Formation of eastern Colorado, the Lance Formation of Wyoming, the Hell Creek Formation of Montana and North and South Dakota, the Frenchman Formation of Saskatchewan, and the Upper Edmonton Formation of Alberta. The known range in latitude of this community is from 39° to 52° N. This community is characterized by the genera *Triceratops*, a horned dinosaur, and *Anatosaurus*, a flat-headed duckbill dinosaur. Regional variations in the relative numbers of dinosaurs are small (summarized by D. A. Russell, 1967, and Sloan, 1969). There are at present no genera of herbivorous dinosaurs common to this community and the *Alamosaurus* community to the south. These two communities were physically separated, at least in part, by mountain ranges in southern Wyoming and central Colorado (Wiemar, 1960).

Both of these communities are descendants of the community first seen in the Coniacian Djadochta Formation of Mongolia and best known from the Campanian Oldman Formation of Alberta (Langston, 1965; D. A. Russell, 1967) and the Judith River Formation of Montana (Sahni, 1972). In many ways the more southerly of the two Maastrichtian communities resembles the Campanian community more closely. Two genera of Campanian hadrosaurs survive into the Maastrichtian only in the *Alamosaurus* community. This in itself suggests that the climate characteristic of the Campanian at 50° N no longer existed at that latitude in the Maastrichtian, but was present only toward the south to a latitude of 39° N. From the mid-Campanian to the Maastrichtian there would seem to be a 10 degree shift of climatic belts towards the equator.

Because of the slow rate of evolution of the late Cretaceous dinosaurs of North America and the extreme differences in dinosaur biofacies from Alberta to Mexico, it is impossible to determine on the basis of internal evidence whether the northern or the southern dinosaurs became extinct first. However, the taxonomic rates of evolution in the replacing community are among the most rapid known. This is the Puercan arctocyonid-hyopsodont-periptychid primitive ungulate community. The species of this community evolved and diversified so rapidly that no two collected samples of it are substantially identical. When these samples are compared, it is possible to arrange them in a temporal sequence, based on evolutionary trends in several lineages and controlled by stratigraphically superposed local faunas. The Puercan, on the basis of radiometric dates and extrapolation of rates of sedimentation, appears to represent a time span of about 1.5 million years. The Puercan mammals deposited immediately above the last dinosaurs in Montana and northern Wyoming at 45° N are phyletically earlier than those of the San Juan Basin of New Mexico, at 36° N. On this basis alone it appears likely, though not proven, that the last dinosaurs in Montana became extinct about one half-million years earlier than those in New Mexico.

Since there are insufficient data available to discuss adequately the general problem of the extinction of the latest Cretaceous dinosaurs throughout the world, I shall restrict myself to the history of extinction of the *Triceratops* community. This is the only dinosaur community for which relatively complete data on community structure and stratigraphic distribution exist. I shall attempt to show that this community became extinct as a result of a climatic change that led to a reduction in productivity and diversity of the primary producers and a consequent disruption of the higher trophic levels.

The *Triceratops* community was a stable one, in which little significant evolution of any component has been shown for the duration of the community. Its composition and disruption can be described in a semi-quantitative way on the basis of field observations in the Hell Creek Formation during the past decade.

The late Cretaceous Bearpaw Shale, Fox Hills Sandstone, and Hell Creek Formation were deposited in the epicontinental sea, which extended from Alberta and Saskatchewan to Texas; and on the bordering coastal plain. The source for these clastic sediments, as well as for those of the Paleocene Fort Union Group, were the Laramide Rocky Mountains, located in western Montana and Wyoming. The prevailing drainage was about 60° east of north, roughly normal to the trend of the continental divide. The sea retreated to the south at least beyond the latitude of southern Colorado by the end of the deposition of the *Triceratops* zone. The basal part of the early Paleocene Tullock and Ludlow formations of the Fort Union Group was then deposited on a broad interior plain which probably drained southward, along the axis of the former Bearpaw-Pierre seaway, to the Gulf of Mexico. The early and mid-Paleocene seaway in which the Cannonball Formation was deposited then transgressed along this axis to a position south of the Canadian border and east of the Montana–North Dakota line. The coastal plain bordering this sea on the west was the site of deposition of the Tullock and Ludlow formations, the Lebo Shale, and the Tongue River Formation.

The Bearpaw, Fox Hills, and Hell Creek formations represent three realms of deposition or lithofacies which are broadly contemporaneous. The Bearpaw Shale is a dark grey, shallow water, marine shale; the Fox Hills Sandstone is a sequence of beach, bar, lagoon, and coastal paludal clastic deposits; and the Hell Creek Formation is a sequence of stream deposits on the coastal plain, consisting of flood plain shales and silts and channel sandstones. As a result of the retreat of the sea and consequent lateral shift of the three principal environments of deposition, the precise age of the contacts between these formations varies considerably from place to place.

In any given area within the northern plains the sequence of formations (and environments) is marine Bearpaw (or upper Pierre) Shale, strandline Fox Hills Sandstone, and the coastal plain Hell Creek Formation. Waage (1961) has demonstrated that in northwestern South Dakota, the Hell Creek Formation grades easterly into the Fox Hills Sandstone and in turn into the marine Pierre Shale. Much the same change can be demonstrated in the region of the Fort Peck Reservoir, Montana, where an eastward stratigraphic climb of the Fox Hills facies can be demonstrated. In a distance of 110 miles (175 km), from the intersection of the Musselshell River and the Missouri River to Poplar, Montana, the Hell Creek Formation thins from 360 feet to 120 feet (110 to 37 m). All of the change in thickness is due to lateral gradation of the Hell Creek Formation into rocks of Fox Hills facies and of Fox Hills Sandstone into Bearpaw Shale. Obvious cycles of aggradation can be seen in all formations except the Bearpaw shale. These cycles begin with an abrupt increase in grain size, coarse sand and clay pebbles in the channels, fine sand or silt on the flood plains. These coarse basal sediments are usually oxidized yellow or red. They are followed by olive silts and clays. Each cycle is terminated by an organic-rich sediment; a "chocolate brown" shale rich in macerated plant fragments in the case of the Hell Creek, or a lignite in the case of the Tullock. Each cycle must represent a sizable span of time, because coniferous stumps two feet (60 cm) in diameter are rooted in the middle shales of several cycles at one locality in the Tullock Formation and are buried by later beds of the same cycles. No unconformity more significant than the usual intraformational unconformities normally present between cycles of aggradation in non-marine sediments can be or has been demonstrated for the entire sequence of Cretaceous and Paleocene rocks in this area. The area seems to have been unusually stable, with a rate of subsidence of basement rocks distinctly less than the rate of sedimentation.

The Z-coalbed, as well as others in the Tullock Formation, was deposited on interfluvial divides as well as in flood plains. It pinches and swells in thickness and shows many local variations in sedimentary facies. Because the lignites occupy the same relative position in the cycles of alluviation as the organic-rich shales of the Hell Creek Formation, some climatic factor must have been involved in determining whether organic-rich shales or lignites were deposited. R. W. Brown (1962) has shown that the lower Fort Union lignites in North Dakota grade laterally into tongues of the Cannonball Formation. These are separated by tongues of non-marine or strandline sediments. The swamps which produced either lignite or organic-rich shales, depending on climate, might best be explained by a high water table, which would accompany the slowest rates of sedimentation and lowest absolute elevations of the land surface. These high water tables would inhibit decomposition of duff and plant tissue and facilitate accumulation of peat and/or organic shales. The lower temperature for the early Paleocene as compared to the latest Cretaceous, suggested on the basis of palaeobotanical evidence by Dorf (1940), would further inhibit decomposition of organic detritus and may be the reason for the differing characters of the organic sediments in the late Cretaceous and early Paleocene of the region.

Norton (Norton, 1963; Norton and Hall, 1967; Hall and Norton, 1967) made a careful study of the palynology of the sediments exposed in the vicinity of Hell Creek. The rocks investigated included those between the upper Bearpaw Shale and the middle of the Lebo Formation. He found 144 species of pollen and spores in 100 samples over a stratigraphic range of 820 feet (250 m). His studies resulted in a detailed zonation of these rocks and the most complete analysis to date of the changes in floral composition occurring across the Cretaceous–Tertiary boundary. The study shows a continuously changing flora with no sharp breaks in stratigraphic distribution. These results will be summarized in several ways to describe floral and climatic changes in this region. It is well known that there is a high positive correlation between floral diversity, expressed in numbers of species, and climatic rigor, particularly temperature. A table of floral diversity expressed as numbers of species of pollen represented in

Table 1. Changes in the percentages of pollen and spores between the Upper Bearpaw and Lower Lebo formations.

Stratigraphic unit	Thickness ft/m	Number of spore and pollen species present	Per cent of species in Lower Hell Creek flora	Per cent of species in Tullock flora
Lower Lebo Formation	189/53	34	18%	42%
Tullock Formation	163/47	59	29%	100%
Upper Hell Creek Formation	120/37	69	59%	76%
Lower Hell Creek Formation	140/43	107	100%	52%
Fox Hills Sandstone (marine)	129/40	37	34%	29%
Upper Bearpaw Shale (marine)	80/25	29	27%	20%

each stratigraphic interval provides a crude indication of climates. Norton showed that the flora typical of that portion of the Hell Creek Formation in which dinosaurs are most frequent was gradually replaced by a very different flora, characteristic of the Tullock Formation, over a stratigraphic interval of about 120 feet (37 m) in the upper part of the Hell Creek Formation. Table 1 shows the percentages of species in the lower Hell Creek and Tullock floras for each stratigraphic interval, and shows how the floras were replaced.

The small taxonomic diversity of the flora in the Bearpaw Shale and the Fox Hills Standstone reflects the marine character of these rocks. The pollen and spores in these sediments were derived from the nearby coastal plain, present in western Montana, and do not represent a local flora but rather a "background" of sporomorphs capable of being transported long distances. The other formations are all non-marine, and quite comparable in their range of depositional environments. The pollen and spores in these sediments were for the most part locally derived.

The pronounced reduction in floral diversity from the lower Hell Creek Formation (late Cretaceous) to the lower Lebo Formation (middle Paleocene) suggests a cooling trend very similar to that proposed by Dorf (1940) on the basis of fossil leaves. The same cooling trend is suggested by the present climatic distribution of the plant taxa represented in these stratigraphic units. Within this stratigraphic sequence, there is a two-fold increase in forms with modern temperate distribution and a three-fold decrease in forms with modern tropical distribution.

The stratigraphic distribution of the pollen and spores further suggests that a warm temperate to sub-tropical rain-forest was the characteristic vegetation of the lower Hell Creek Formation in this region and that it was gradually replaced, during the deposition of the upper Hell Creek Formation, by a cooler temperate, mixed coniferous and deciduous forest.

A suggested restoration of the *Triceratops* community, looking rather different from previous restorations, is shown in Fig. 1. Silicified coniferous wood is rarely if ever found in the Hell Creek Formation but is relatively common in the Tullock Formation. At least five successive petrified forests, represented by silicified stumps of *Metasequoia*, are present in the lower half of the Tullock Formation in Section 33 T24N, R44E, in McCone County, Montana. The stumps in these forests have a diameter (breast high) of about 2 to 4 feet (.6 to 1.3 m), suggesting that the Tullock forests had crown heights of some 60 to 120 feet (19 to 37 m). The stumps are rooted in the middle shales of the alluviation cycles of the Tullock. Nearby, in Section 35, T24N, R44E, a series of 32 *Metasequoia* stumps is exposed, all rooted in the uppermost bed of the W-coalseam of the lower Tullock

Fig. 1 Restoration of the *Triceratops* community from the Hell Creek Formation of northeastern Montana.

Fig. 2 Restoration of the *Taeniolabis* community from the petrified forest developed on the
W-coalbed of the Tullock Formation of McCone County, Montana, same scale as
Fig. 1.

Formation. These stumps vary from 1 foot to 6 feet (.3 to 2 m) in diameter
(breast high). Fig. 2 is a restoration of part of this forest, derived from a map
of the stumps. The height of the trees was calculated from stump diameters, based
on the proportions of modern *Metasequoia glyptostroboides*. This site is 8 miles
(13 km) from, and approximately contemporaneous with, the Purgatory Hill
site that has yielded mammals and leaves (Sloan and Van Valen, 1965; Van Valen
and Sloan, 1965; Shoemaker, 1966). The largest known multituberculate,
Taeniolabis taoensis, which is roughly the size of a paca (*Cuniculus*) or small
capybara (*Hydrochoerus*); and *Eoconodon* sp., an omnivorous ungulate roughly
the size of a small sheep, were the largest mammals and terrestrial herbivores in
this community.

  The complexity of a community is measured by the number of its interspecific
interactions, such as parasitism, predation, mutualism, and competition. Without
inquiring into the exact nature of each of these relationships, this complexity can
be estimated as the square of the number of species of organisms present in the

community. This formula is derived from the supposition that these interrelationships (expressed quantitatively in terms of changes in biomass, populations, or
selection coefficients) form an $n \times n$ matrix where n is the total number of species.
Not all the interrelationships are significant. Measuring these entries is more
difficult than conceiving their existence. It seems reasonable to suppose that the
greater the complexity of a community, the greater the inherent stability.

The observation has frequently been made that the diversity of species and
complexity of community structure increase towards the equator (Odum, 1959;
Darlington, 1957; Cousminer, 1961; Fischer, 1960). The complexity is, in the
long run, essentially determined by the major physical variables of the environment, principally by the interrelated suite of temperature, precipitation, altitude,
and latitude.

If it is difficult to achieve insight into the structure of a modern community of
any great complexity, it is doubly so in the case of fossil communities. There are
no fossil communities for which even species lists are substantially complete. A
description of the collecting techniques used to obtain the present data will
suggest the problems. An estimate of the flora can be made on the basis of
megafossils, principally leaves, or on the basis of the pollen and spores preserved
in the sediments. These two types of collections have different biases. The leaf
collections have been reported by R. W. Brown (1939, 1962), Dorf (1942), and
Shoemaker (1966). Each site investigated by these authors produced a limited
variety of leaves, essentially those that originally grew a very short distance from
the site of deposition. An adequate flora can be recovered only by very extended
collecting, and the number of species can usually be increased by further collecting. This number might reasonably be expected to approach a limit exponentially
with continued collection, but does not appear to do so. This preservation bias
would mean underrepresentation of plants far removed from watercourses, on
interfluvial divides.

This bias in the preservation of leaves is counteracted to a considerable extent
by the very different bias in the deposition of pollen and spores. The latter bias is
in favour of wind-pollinated plants and against insect-pollinated plants. Of the
two sorts of collections, the pollen floras are more varied than the leaf floras and
afford closer estimates of the total flora, but they need to be supplemented by
floras based on foliage for fuller understanding. A comparison of the pollen floras
described by Norton (1963) and Stanley (1965) with the leaf floras recorded
by R. W. Brown (1962), Dorf (1942), and Shoemaker (1966) indicates that
the angiosperms are grossly underrepresented in the pollen floras (Table 2).
Monocotyledons are underrepresented by about 200 per cent, and dicotyledons
by about 300 per cent.

There are adequate samples of the freshwater Mollusca, and their collection
poses no special problems. The vertebrates, other than birds, are adequately
sampled by a combination of surface prospecting for skulls and articulated
skeletons of the larger vertebrates (B. Brown, 1907), and the recently exploited
washing techniques (McKenna, 1960; Clemens, 1963, 1966, 1973; Estes, 1963;
Estes and Berberian, 1969; Sloan and Van Valen, 1965; Van Valen and Sloan,
1965).

The totally unsampled portion of these communities includes the insects, which

Table 2. Species in floras of the Lower Fort Union Group.

Flora	Tullock and Lebo formations – Pollen flora (Norton, 1963)	Lower Fort Union Group – Leaf flora (R. W. Brown, 1962)
Bryophyta	1	2
Lycopsida	0	3
Sphenopsida	1	1
Filicales	14	13
Gymnospermae	18	9
Monocotyledonae	2	7
Dicotyledonae	23	63
	—	—
	59	93

were surely of major importance in food webs. However, they are not preserved, nor are the various equally important "worms", algae, and soil flora. The true complexity of these fossil communities, then, is probably 10 to 100 times greater than the estimates based on the known record.

To determine what environmental changes were taking place concurrent with the extinction, changes in community type and structure must be estimated. One such estimate is based on the climate of the area where the maximum number of modern relatives (preferably descendents) of taxa in the fossil communities now occurs. An appropriate estimate of an index of diversity or of the complexity of the community and its changes might then be compared with ecological estimates based on individual taxa. Estes (1964) proposed, on the basis of the lower vertebrates other than dinosaurs, that the southeastern part of the United States might be an appropriate analogue for the Lance community.

The late Cretaceous and Paleocene communities of the North American coastal plain can be divided into subcommunities that had very different, more or less independent histories. One such subcommunity is the stream and stream-bank community first described in detail by Estes (1964) from the Lance Formation of Wyoming. It is composed of a characteristic suite of fish, neotenic salamanders, frogs, turtles, crocodilians, and *Champsosaurus*. There are few changes in its composition, except by slow evolution at the specific level, from the mid-Cretaceous to the early Eocene (Estes, 1963), and it is known from Mongolia, from North America from Texas to Canada, and from the Paris Basin. The relative proportions of its elements (and thus the trophic web) remain relatively constant until the introduction of abundant freshwater teleosts in the Eocene. Whatever changes that disrupted other community structures at the end of the Cretaceous did not alter this community.

The other vertebrate subcommunities present in the coastal plain of this area are the *Triceratops* community of dinosaurs with one suite of mammals (Estes, 1964; Clemens, 1960, 1963, 1966), and the *Protungulatum-Stygimys* community with a distinctly different suite of mammals (Sloan and Van Valen, 1965). These show a replacing relationship. Fig. 3 illustrates the trophic levels of these communities.

The following data on the distribution and relative abundance of dinosaurs can be compared with those of D. A. Russell (1967).

The only species of *Triceratops* present in the Hell Creek Formation is *T. prorsus* (including *T. brevirostris* and *T. maximus*), and it is significantly larger than at Lance Creek. The smallest adult *Triceratops* skull collected from the Hell Creek Formation is the size of the average Lance Creek skull, while the largest Hell Creek skulls are about 1 foot (.3 m) longer than the largest known Lance Creek skull. This difference agrees with Bergman's rule of the relationship between latitude and the size of warm-blooded animals. Colbert, Cowles, and Bogert (1946) have previously suggested that large dinosaurs were functionally homoeothermic. *Triceratops* is approximately 10 times as numerous as the next most frequent dinosaur, *Anatosaurus copei*. These two dinosaurs fulfil the role of large herbivores in this community. Although Barnum Brown of the American Museum of Natural History collected an ankylosaur (*Paleoscincus* sp.) from the lower Hell Creek Formation in 1906 and 1908, this suborder is quite rare; our parties have not found any scraps in seven years of field work. The small dinosaurs include the ornithischian *Thescelosaurus neglectus* and the coelurosaurs *Paronychodon lacustris* and *Ornithomimus altus*. These small dinosaurs are approximately equally numerous, in roughly one-tenth the abundance of *Anatosaurus*; each is roughly 10 times as abundant as the medium-sized carnosaur, *Gorgosaurus lancensis*, which in turn is 10 times as abundant as the largest carnosaur, *Tyrannosaurus rex*.

The top carnivore, *Tyrannosaurus rex*, has been found as articulated skeletons only in the lower middle Hell Creek Formation (field notes of Barnum Brown, 1902 and 1908). The stratigraphically youngest specimen known, now in the Los Angeles County Museum, was found 57 feet (17 m) below the Z-coalbed.

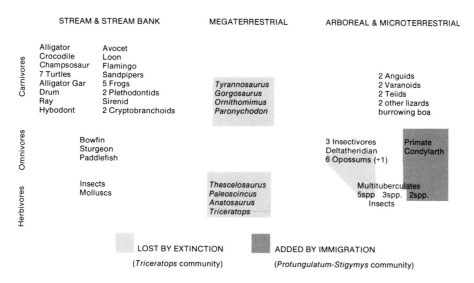

Fig. 3 Trophic levels in Montana showing changes from the *Triceratops* community to the *Protungulatum-Stygimys* community (after Sloan, 1969).

The peak abundance of dinosaur remains in the formation is reached about 120 to 100 feet (33–27 m) below the Z-coalbed. This is also the horizon at which the transition from the flora of the lower Hell Creek Formation to that of the Tullock Formation begins. Dinosaur remains in the lower and middle Hell Creek Formation are roughly 10 times as frequent per area of outcrop as those above this horizon. The remaining dinosaurs occur with the same relative frequencies up to the Z-coalbed at the base of the Tullock Formation.

Articulated skeletons of *Triceratops* and *Anatosaurus* were collected by Barnum Brown 20 feet (6 m) below the Z-coalbed in McCone County, Montana, and 3 skulls of *Triceratops* were collected by James Jensen of Brigham Young University some 35 feet (11 m) below the Z-coalbed in Garfield County, Montana.

The fossil localities in the Hell Creek Formation that yield mammals show two distinctly different facies (Sloan and Van Valen, 1965). The first facies yields members of the *Triceratops* community and is directly comparable to that of the Lance Formation of Wyoming. Mammals were first collected from the Hell Creek Formation by Barnum Brown in 1906 at two localities in the upper Hell Creek Formation, one on Crooked Creek, the other on Gilbert Creek. Brown washed matrix through a burlap sack to concentrate the fossils. Simpson (1927) published a short note on Brown's Crooked Creek collection, but the Gilbert Creek collection has not been found or studied. This facies is present at all of the mammal localities except Bug Creek Anthills, Bug Creek West, and Harbicht Hill. The mammals alone did not provide the balanced superstructure of a terrestrial community, but were complemented by reptilian herbivores and carnivores (Estes, 1964). The samples of mammals from these localities are all small but, individually and in total, are indistinguishable from the mammalian portion of the *Triceratops* community described by Clemens (1960, 1961, 1963, 1966, 1973) from the Lance Formation of Lance Creek Wyoming, save for the replacement of *Meniscoesus robustus* by the larger *M. borealis*. A comparison of the large quarry samples from Lance Creek, Wyoming (UCMPV-5620 and V-5711) with those from Bug Creek Anthills shows that, with the exception of the three species of Ectypodontidae, the species from the Lance facies occur in similar relative numbers in both collections. This, then, represents the element of community in the strictest sense. The dinosaurs form an equally important part of this community and show similar changes in abundance. Most of the mammals of the *Triceratops* community apparently became locally extinct in the upper Hell Creek Formation of the Fort Peck region.

The samples from the three localities of the Bug Creek facies are dominated by quite a different assemblage. If the elements of the Lance facies, i.e., the non-ectypodont multituberculates (small herbivores), the marsupials (small carnivores), the ornithischians, the carnosaurs, and the large coelurosaurs are subtracted, there remains a community dominated by small mammals, taxonomically similar at the ordinal or family level to those of the Puercan age.

Bug Creek Anthills, with the channel-base 80 feet and channel-top 56 feet (25 and 17 m) below the Z-coalbed, produces fossil mammals mainly of the *Protungulatum-Stygimys* community. Three members of the *Triceratops* community, *Mesodma formosa*, *Mesodma thompsoni*, and *Cimexomys minor*, are also members, even dominant ones, of the *Protungulatum-Stygimys* community.

The leptictids *Gypsonictops hypoconus* and *G. petersoni* also occur in both. All of the other mammalian members of the *Triceratops* community are present at Bug Creek Anthills in about the same proportions of minimum numbers of individuals relative to each other as at Lance Creek and together total about five per cent of the mammalian fossils. It is obvious, then, that these communities coexisted and that the *Triceratops* community existed intact and little altered some unknown distance upstream (Shotwell, 1955, 1958, 1963). Bug Creek West, like Bug Creek Anthills, has the three species of ectypodonts in common with the *Triceratops* community and, like Bug Creek Anthills, has yielded a number of isolated teeth of other species from the *Triceratops* community. These isolated teeth again total five per cent of the minimum number of individuals present. The Brownie Butte locality, 41 miles (66 km) west (upstream), exhibits the *Triceratops* facies and is just below the same horizon as Bug Creek Anthills, with the top of the channel 88 feet (27 m) below the Z-coalbed. A nearby higher locality in the Hell Creek Formation that produced specimens of *Catopsalis joyneri* and *Protungulatum* sp. "S" was discovered in 1964 and rediscovered in 1973. The latest Hell Creek locality, Harbicht Hill, where the base of the channel is 40 feet (12 m) below the Z-coalbed and the top of the channel is 10 feet (3 m) below the Z-coalbed, shows only a few survivors of the *Triceratops* community. Only one mammalian species, *Meniscoessus borealis*, is restricted to the *Triceratops* community and occurs as a member of a distal community from this locality. There was some doubt, in 1938, about the origin of these *Meniscoessus* specimens. Collections made for the present study certainly show that the species does occur at this locality. Since the rest of the specimens known in 1938 were all from the *Protungulatum-Stygimys* community, it is not surprising the balance of the 1938 collection was thought to be early Paleocene in age. *Mesodma thompsoni* is a member of this community, but *M. formosa* seems quite surely to be missing and *Cimexomys minor* has been replaced by its larger undescribed descendant. The two species of *Gypsonictops* also occur here as members of the *Protungulatum-Stygimys* community. Thus it seems likely that by this time, the *Triceratops* community had been either almost disrupted by extinction of its component species or restricted to an area much farther upstream than before. In view of the number of dinosaur teeth collected at this outcrop, the former seems more likely. This likelihood is increased by the observed presence of articulated dinosaurs at still higher levels in the formation (American Museum of National History 5033, skeleton of *Triceratops prorsus*, part of composite mount, from 20 feet (6 m) below the Z-coalbed, NE 1/4 T24N, R43E, notes of Barnum Brown). Table 3 shows the manner in which the mammals of the *Triceratops* community disappear.

On the basis of quantitative samples collected by washing at Bushy Tail Blowout (University of California Museum of Paleontology, loc. V-5711), assuming that the average life span of *Triceratops* and *Anatosaurus* was 25 years and that their teeth were completely replaced (including those in the dental magazines) once a year, calculations indicate that these herbivores were outnumbered by the mammals by about 300 to one in this community. This result seems to be reasonable in view of what is known of the relationships between size and abundance in modern communities (Odum, 1959).

Table 3. Stratigraphic distribution and relationships of mammals
from the *Triceratops* community.

Species	80 ft. (24 m) Bug Creek Anthills and below	60 ft. (18 m) Bug Creek West and Brownie Butte	40 ft. (12 m) Harbricht Hill	Also members of *Protungulatum-Stygimys* community	Has Paleocene descendant
Cimexomys minor	+	+	−	+	+
Mesodma formosa	+	+	−	+	+
Mesodma thompsoni	+	+	+	+	+
Cimolodon nitidus	+	+	−	−	?
Cimolomys gracilis	+	−	−	−	−
Essonodon browni	+	−	−	−	−
Meniscoessus borealis	+	+	+	−	−
Didelphodon vorax	+	−	−	−	−
Alphadon marshi	+	+	−	−	+
Pediomys elegans	+	+	−	?	−
Pediomys cooki	+	+	−	−	−
Pediomys hatcheri	+	+	−	−	−
Glasbius intricatus	+	+	−	−	−
Cimolestes incisus	+	−	−	+	+
Gypsonictops hypoconus	+	+	+	+	?
Gypsonictops petersoni	+	+	+	+	?

A comparable calculation for the *Protungulatum-Stygimys* community of the upper Hell Creek Formation (based on the samples from Bug Creek Anthills) shows that the proportion there was roughly 8,000 mammals per dinosaur. These calculations suggest that the rarity of Cretaceous mammals in most collections is a result of preservation bias rather than low population density.

While the *Triceratops* community was undergoing disruption, the *Protungulatum-Stygimys* community had begun to diversify. One new species of multituberculate (*Cimexomys* sp.), four new species of *Protungulatum* (P. sp. "O", sp. "H", sp. "S", and sp. "E", Sloan and Van Valen, 1965), and a primate (*Purgatorius ceratops*, Van Valen and Sloan, 1965) evolved in (or, less likely, migrated into) this community during the time represented by the three localities exhibiting this community.

It would seem apparent that the *Protungulatum-Stygimys* community is an immigrant fauna that replaced the *Triceratops* community as a result of some physical change in the environment. The direction of immigration is of some interest. Faunas of Lancian age, containing mammals, are known from Alberta, eastern Montana, South Dakota, eastern and western Wyoming (Polecat Bench), Utah and New Mexico. Of these faunas only those from Garfield and McCone County, Montana, have the *Protungulatum-Stygimys* community or pre-Puercan elements.

Some plausible ancestors for the placental genera *Procerberus* and (less likely) *Protungulatum* occur in the Maastrichtian (Lancian) Scollard (Upper Edmonton) Formation of central Alberta (Lillegraven, 1969). The precise ancestors of *Stygimys* and *Catopsalis*, the most herbivorous of the multituberculates of the invading community, occur in the Coniacian or Santonian Djadochta and the late Campanian Nemegt formations of Mongolia (Kielan-Jaworowska, pers. comm., 1971; and observation of her specimens). No members of their families are known from North America prior to the Lancian, save for a scrap or two of eucosmodontid incisors from Baja California of undetermined age.

The Bearpaw-Pierre Sea was located to the east, effectively blocking direct travel by land mammals from that direction. To the west lay the Laramide Mountains and the California Trough, which would also serve as barriers. An open corridor from what is now the Canadian Arctic plains was available, and this is also the principal direction from which known ancestors could have immigrated. The immigration of the Bug Creek community would then seem to be a range extension of a northerly or temperate coastal plain community.

At this point it would be fruitful to review a list of various *ad hoc* hypotheses that have been proposed to explain the extinction of dinosaurs at the end of the Cretaceous and see which, if any, might apply to this particular dinosaur community. Colbert (1961) has supplied such a list and these hypotheses will be reviewed, although not in the order in which he presented them. None of them is original with Colbert and he shows no special preference for any one of them.

The "racial senescence" theory should perhaps be dismissed first. Excessive gigantism, numerous spines, and bizarre shapes have been regarded as symptoms of a mysterious phylogenetic old age and indications of impending extinction. *Triceratops* and *Tyrannosaurus* are among the largest genera of their respective families but are not particularly bizarre or unusual; they seem remarkably well adapted for the roles they play in the community. None of the other dinosaurs in the community is in the least unusual or among the largest of its family. Finally, it is most difficult to find any good examples of "racial senescence" in any other episodes of extinction or to find any hint of a momentum-effect in the processes of evolution as presently understood.

Colbert has discredited the hypotheses of a loss in viability of the late Cretaceous dinosaur eggs and of an increase in the concentration of atmospheric oxygen, and has also discussed adequately the hypothesis involving epidemics. I might add that postulating an epidemic to account for the extinction of the dinosaurs in this community would require totally unreasonable rates of sedimentation, considering their progressive decline in abundance in 120 feet (37 m) of sediments.

Clemens (1960) and Colbert (1961) have both discussed and dismissed the twin hypotheses of competition between mammals and dinosaurs and of mammalian predation on dinosaur eggs. The mammals were an integral part of the *Triceratops* community throughout its existence, and very similar mammals were present in the preceding and ancestral communities of the Oldman, Edmonton, and Judith River formations. It might legitimately be asked whether or not the mammal members of the *Protungulatum-Stygimys* community might have been more efficient competitors of the dinosaurs than those of the *Triceratops* community. Since the biggest species of mammal in the latest Cretaceous sample of the *Protungulatum-Stygimys* community, *Protungulatum* sp. "H", is approximately raccoon-sized, this is equivalent to expecting a community of mice, squirrels, woodchucks, and raccoons to cause the extinction of a community of elephants, rhinoceros, and lions. It seems most unlikely indeed, although not completely impossible.

Cowles (in Colbert *et al.*, 1946) has proposed the hypothesis that the extinction of dinosaurs was brought about by an increase in average world temperatures at the end of the Cretaceous. He offered, in support of his hypothesis, some ingenious experiments in comparative physiology. If such an increase were of very short duration (a few years or so) we would not be in a position to dispute it. However, considering that the number of dinosaurs in the Fort Peck region is observed to have declined more or less steadily through an interval in which the palaeobotanical evidence suggests a climate gradually becoming cooler, Cowles' hypothesis, while it may be valid for some other cases, certainly does not seem to apply here.

The two remaining hypotheses on Colbert's list seem to be much more applicable to our problem of the extinction of the *Triceratops* community. The first of these is that the Laramide Revolution, or events associated with it, brought about a world-wide lowering of average temperatures. Whether or not this was world-wide, the palaeobotanical evidence definitely indicates an increasingly temperate climate for the northern Great Plains Region. The most obvious reason for this drop in temperature is the retreat of the Bearpaw-Pierre Sea. This event was one of the aspects of the Laramide Revolution and would have changed the climate of eastern Colorado, Wyoming, Montana, Alberta, and Saskatchewan from a maritime subtropical or warm temperate climate with mild winters and warm humid summers to a more continental climate of cold winters and hotter summers. Since this sea retreated from north to south, the climatic effect would have been felt earlier in the north and would have advanced progressively to the south, becoming more intense at any latitude as the sea grew more distant. This increased seasonality and generally cooler temperatures might be expected to have had more effect on the terrestrial reptiles than on the aquatic forms. This direct effect is, however, probably less important than some of the secondary effects of this climatic change.

The only remaining hypothesis on Colbert's list is "that changes brought about by the Laramide Revolution resulted in changes of food supply. The herbivorous dinosaurs as a result died out and with their passing the carnivores also disappeared". He goes on to state, "The dinosaurs of late Cretaceous times did live in modern forests of a modern aspect, so if there was a change in food supply that affected the plant-eating dinosaurs it was too subtle to be registered in the geologic record". It was possible for Colbert to say this in 1961 when the only palaeobotanical evidence bearing on the problem were Dorf's (1940, 1942) studies on small collections and Roland Brown's unpublished Fort Union flora. Since Colbert wrote, however, a number of palynological studies bearing on the question have been completed (Anderson, 1960; Stanley, 1965; Leffingwell, 1970; Srivastava, 1967; and especially Norton, 1963). Of these, Norton's is most useful because it is directly tied to the same measured sections as my observations on fossil vertebrates. The other studies essentially extend the area for which conclusions can be drawn. The changes in the food supply of the herbivores, including the herbivorous dinosaurs, have now been documented more completely and with closer stratigraphic sampling. There is a 45 per cent drop in floral diversity from the horizons in which *Triceratops* is best developed to the first horizon in which it is absent; there is also a loss of 71 per cent of the plant taxa represented in the horizons of peak or optimum development of the dinosaur community. This can no longer be called subtle.

This radical change in flora can be expected to alter the upper trophic levels as well. The moors and bogs in which the Tullock lignites accumulated and the redwood forests which alternated with them could hardly have supplied the kinds of vegetation capable of supporting a rain-forest fauna such as that of the *Triceratops* community.

The *Triceratops* community was confined to a coastal plain with a warm temperate to subtropical rain-forest, bordered on the east by the Bearpaw-Pierre Sea, on the southwest by the Front Range and the southern Wyoming Ranges, on the west by the Continental Divide in western Wyoming, Montana, and Alberta, and on the north by a temperate coniferous forest which was advancing southward following the retreat of the Bearpaw-Pierre Sea. Under these circumstances, it is not hard to see why this endemic community should have failed when the seaway retreated far enough southeast of eastern Colorado to result in the complete alteration of the climatic and floristic character of the region.

The reasons for the extinction of each dinosaur community must be deduced separately on the basis of detailed stratigraphic palaeontology and palaeoecology. It may well be that hypotheses rejected for the extinction of the dinosaurs of the *Triceratops* community might still be valid for dinosaur communities of other continents.

In short, the process of extinction of dinosaur communities in North America probably began in what were temperate regions and may have taken a significant time to advance to the south.

The reduction in primary production in coastal plain and pelagic communities at the end of the Cretaceous is thus well established on the basis of the fossil record, and the evidence agrees with the theoretical analysis of Axelrod and Bailey (1968). This reduction, of itself, is sufficient explain the late Cretaceous spectrum of animal extinctions. The observed rates of extinction (species per million years) are by no means as severe as those of the late Pleistocene, but rather smaller by factors of 10 to 1000.

Literature Cited

ANDERSON, R. Y.
   1960   Cretaceous-Tertiary palynology, eastern side of the San Juan Basin, New Mexico. New Mexico State Bureau of Mines and Mineral Resources, Memoir 6: 1–59.

AXELROD, D. I. and H. P. BAILEY
   1968   Cretaceous dinosaur extinction. Evolution, 22: 595–611.

BRAMLETTE, M. N.
   1965   Massive extinctions in biota at the end of Mesozoic Time. Science, 148: 1696–1699.

BROWN, B.
 1907   The Hell Creek Beds of the upper Cretaceous of Montana: their relationship to
         contiguous deposits with faunal and floral lists and a discussion of their correla-
         tion. Am. Mus. Nat. Hist. Bull., 25: 355–380.

BROWN, R. W.
 1939   Fossil plants from the Colgate Member of the Fox Hills Sandstone and adjacent
         strata. U.S. Geol. Surv. Prof. Pap. 189–I: 239–275.
 1962   Paleocene flora of the Rocky Mountains and Great Plains. U.S. Geol. Surv. Prof.
         Pap. 375: 1–119.

CLEMENS, W. A.
 1960   Stratigraphy of the Type Lance Formation. Internat. Geol. Cong., Rept. Twenty-
         first Session, pt. V: 7–13.
 1961   A late Cretaceous mammal from Dragon Canyon, Utah. J. Paleo. 33: 578–579.
 1963   Fossil mammals of the Type Lance Formation, Wyoming, Part I. Introduction and
         multituberculata. Univ. Calif. Publ. Geol. Sci., 48: 1–105.
 1966   Fossil mammals of the Type Lance Formation, Wyoming, Part II. Marsupialia.
         Univ. Calif. Publ. Geol. Sci., 62: 1–122.
 1973   Fossil mammals of the Type Lance Formation, Wyoming, Part III. Eutheria and
         summary. Univ. Calif. Publ. Geol. Sci., 94: 1–102.

COLBERT, E. H.
 1961   Dinosaurs, their discovery and their world. New York, Dutton & Co., 1–300.

COLBERT, E. H., R. B. COWLES and C. M. BOGERT
 1946   Temperature tolerances in the American alligator and their bearing on the habits,
         evolution and extinction of the dinosaurs. Amer. Mus. Nat. Hist. Bull., 86: 327–
         374.

COUSMINER, H. L.
 1961   Palynology, paleofloras and paleoenvironments. Micropaleontology, 7: 365–368.

DARLINGTON, P. J.
 1957   Zoogeography, the geographic distribution of animals. New York, John Wiley and
         Sons, Inc.: 1–675.

DORF, E.
 1940   Relationship between floras of the Type Lance and Fort Union Formations. Geol.
         Soc. Am. Bull., 51: 213–235.
 1942   Upper Cretaceous floras of the Rocky Mountain Region. Publ. Carneg. Inst. Wash.,
         508: 1–168.

ESTES, R.
 1963   Late Cretaceous and early Cenozoic lower vertebrate faunas in North America
         (abstract). Geol. Soc. Am., Spec. Pap. 73: 1–146.
 1964   Fossil vertebrates from the late Cretaceous Lance Formation, eastern Wyoming.
         Univ. Calif. Publ. Geol. Sci., 49: 1–187.

ESTES, R. and P. BERBERIAN
 1969   Paleoecology of a late Cretaceous vertebrate community from Montana. Mus.
         Comp. Zool., Harvard Univ. Breviora #343.

FISCHER, A. G.
 1960   Latitudinal variations in organic diversity. Evolution, 14: 64–81.

HALL, J. W. and J. J. NORTON
 1967   Palynological evidence of floristic change across the Cretaceous-Tertiary boundary
         in eastern Montana. Palaeogeogr., Palaeoclimatol., Palaeoecol., 3: 121–131.

JEPSEN, G. L.
  1964   Riddles of the terrible lizards. Amer. Scientist, 52: 227–246.

LANGSTON, W. JR.
  1965   Pre-Cenozoic vertebrate paleontology in Alberta: its past and future. Vertebrate
          paleontology in Alberta. Rept. of a conference held at the University of Alberta
          Aug. 29–Sept. 3, 1963. Edmonton, Univ. Alberta Press: 1–9.

LEFFINGWELL, H.. A.
  1970   Palynology of the Lance and Fort Union Formations of the Type Lance area,
          Wyoming. Geol. Soc. Am., Spec. Pap. 127: 1–64.

LILLEGRAVEN, J.
  1969   Latest Cretaceous mammals of upper part of Edmonton Formation of Alberta,
          Canada, and review of marsupial-placental dichotomy in mammalian evolution.
          Kansas Univ. Paleont. Contrib. Art. 50: 1–122.

LIPPS, J. H.
  1970   Plankton evolution. Evolution, 24: 1–22.

LULL, R. S.
  1933   A revision of the Ceratopsia or horned dinosaurs. Peabody Mus. Nat. Hist. Mem.,
          3 (pt 3): 1–175.

LULL, R. S. and N. E. WRIGHT
  1942   Hadrosaurian dinosaurs of North America. Geol. Soc. Amer. Sp. Pap. 40: 1–242.

MCKENNA, M. C.
  1960   Fossil mammalia from the early Wasatchian Four Mile Fauna, Eocene of Colo-
          rado. Univ. Calif. Publ. Geol. Sci., 37(1): 1–130.

NEWELL, N. D.
  1962   Paleontological gaps and geochronology. J. Paleo., 36: 592–610.

NORTON, N. J.
  1963   Palynology of the upper Cretaceous and lower Tertiary in the Type Locality of
          the Hell Creek Formation. Ph.D. thesis. Minneapolis, Minn., Univ. Minn.: 175 pp.

NORTON, N. J. and J. W. HALL
  1967   Guide Sporomorphae in the upper Cretaceous–lower Tertiary of eastern Montana.
          (U.S.A.): Rev. Palaeobotan. Palynol., 2: 99–110.

ODUM, E. P.
  1959   Fundamentals of ecology. Rev. ed. Philadelphia, W. B. Saunders Co.: 1–546.

OLTZ, D. F.
  1968   Numerical analysis of palynological data from Cretaceous and early Tertiary
          sediments in east central Montana. Paleontographica Abt. B, Bd. 128: 1–166.

OSTROM, J. H.
  1961   Cranial morphology of the hadrosaurian dinosaurs of North America. Amer.
          Mus. Nat. Hist. Bull., 122: 33–186.

RUSSELL, D. A.
  1967   A census of dinosaur specimens collected in western Canada. Natl. Mus. Canada,
          Nat. Hist. Pap. 36: 1–13.

RUSSELL, L. S.
  1973   Geological evidence on the extinction of some large terrestrial vertebrates. Can. J.
          Earth Sci., 10: 140–145.

SAHNI, A.
   1972   The vertebrate fauna of the Judith River Formation, Montana. Bull. Amer. Mus.
           Nat. Hist., 147(6): 321–412.

SHOEMAKER, R. E.
   1966   Fossil leaves of the Hell Creek and Tullock Formation of eastern Montana.
           Paleontographica Abt. B, Bd. 119: 54–75.

SHOTWELL, J. A.
   1955   An approach to the paleoecology of mammals. Ecology, 36: 327–337.
   1958   Intercommunity relationships in Hemphillian (mid-Pliocene) mammals. Ecology,
           39: 271–282.
   1963   The Juntura Basin studies in earth history and paleoecology. Am. Philos. Soc.
           Trans., 53 (pt. 1): 1–77.

SIMPSON, G. G.
   1927   Mammalian fauna of the Hell Creek Formation of Montana. Amer. Mus. Nat.
           Hist. Novitates, 267: 1–7.

SLOAN, R. E.
   1964   Paleoecology of the Cretaceous-Tertiary transition in Montana (Abstract). Science,
           146: 430.
   1969   Cretaceous and Paleocene terrestrial communities of western North America.
           Proceedings of the North Am. Paleont. Conv., Pt. E: 427–453.

SLOAN, R. E. and L. VAN VALEN
   1965   Cretaceous mammals from Montana. Science, 148: 220–227.

SRIVASTAVA, S. K.
   1967   Palynology of late Cretaceous mammal-beds, Scollard, Alberta (Canada). Palaeo-
           geogr., Palaeoclimatol., Palaeoecol., 3: 133–150.

STANLEY, E. A.
   1965   Upper Cretaceous and Paleocene plant microfossils and Paleocene dinoflagellates
           and hystricosphaerids from northwestern South Dakota. Bull. Amer. Paleo.,
           49(222): 179–384.

TAPPAN, H.
   1968   Primary production, isotopes, extinctions and the atmosphere. Palaeogeogr.,
           Palaeoclimatol., Palaeoecol., 4: 187–210.

VAN VALEN, L. and R. E. SLOAN
   1965   The earliest primates. Science, 130: 743–745.

WEIMER, R. J.
   1960   Upper Cretaceous stratigraphy, Rocky Mountain area. Am. Assoc. Petrol. Geol.
           Bull., 44: 1–20.

Peratherium
(Marsupialia: Didelphidae)
from the Oligocene and
Miocene of South Dakota

Morton Green and James E. Martin
Museum of Geology, South Dakota School of Mines and
Technology, Rapid City, South Dakota.

Abstract

The range of the didelphid genus *Peratherium* in the Great Plains extends from the middle Chadronian to the early Hemingfordian. *Herpetotherium* Cope 1873 is considered a synonym of *Peratherium* Aymard 1850. *P. youngi* McGrew 1937 and *P. spindleri* Macdonald 1963 may be synonyms of *P. fugax* (Cope) 1873, but the evidence is inconclusive because the available samples are too small. A possible reduction in molar size may occur between the Oligocene and Miocene populations. Lack of materials prevents absolute demonstration of a chronocline. A provisional faunal list for the Black Bear Quarry II local fauna from the Rosebud Formation is given.

Introduction

This paper is one of a series of studies from the Museum of Geology, South Dakota School of Mines and Technology, dealing with micro-vertebrates from the Oligocene through early Pliocene sequence in South Dakota. These publications provide information from a stratigraphic sequence located in a limited geographic area. This work is part of a study of the micro-mammals from the Black Bear Quarry II local fauna of the Rosebud Formation (early Hemingfordian) in Bennett County, South Dakota. The locality is described by Green (1972).

The most recent "review" of North American Oligocene species of the genus *Peratherium* Aymard is by Hough (1961). Hough, citing Allen (1900) and Tate (1933), begins by discussing the great variation in Recent species of didelphids.

In the systematic portion she contends that minute differences of the dentition are diagnostic for extinct genera and species, particularly the latter.

Also, confusion has been compounded by the repetition of a statement by Cope (e.g. 1884) that the North American species referred to *Peratherium* lack an inflected angle in the mandible and belong to a different genus. This is an odd error since an inflected angle is generally considered an ordinal character of the Marsupialia. Lavocat (1952) questioned the use of the name *Peratherium* for American species, quoting Cope from Scott and Jepsen (1941). (Lavocat's paper is usually cited as 1951, the date printed on the cover and title page. However, the final page suggests that the actual publication date was March 5, 1952.) Hough (1961) then cited Lavocat, completed the circle and resurrected Cope's name, *Herpetotherium*, on this basis. W. von Koenigswald (1970, p. 32) in an important revision of European species of the genus cites and follows Lavocat. Hough's paper is not cited in the text but is listed in von Koenigswald's references. Correction of this error can do much toward clarifying the taxonomic status of some of these species. The fact of the matter is that the species that have been referred to *Peratherium* do have an inflected angle but in most specimens it is broken off. Moreover, Hough recognized some North American specimens as validly belonging to *Peratherium* to the extent of describing a new species. This was evidently overlooked by von Koenigswald.

Simpson (1968) holds that the lower molars of *Peratherium* remained virtually unchanged from its first known occurrence in the Eocene (Greybullian). He further states that the upper molars "may be more distinctive". However, attempts to show that the upper molars are diagnostic have not met with concrete results in differentiating Oligocene and Miocene species.

The result of our study indicates that the Oligocene and Miocene didelphids are probably more variable than the Cretaceous forms.

Specimens of *Peratherium* from the Big Badlands of South Dakota have never been determined specifically in a published record. Clark (1937) cites the genus as occurring in the Upper member (Peanut Peak member) of the Chadron Formation and Macdonald (*in* Bump, 1951) lists it in the Orella member (Scenic member) of the Brule Formation.

Measurements and Terminology

Measurements in mm were made with dial calipers under a binocular microscope. These measurements are comparable to those of Clemens (1966, p. 4). Heavily worn teeth were not included. Statistical comparisons are not conclusive because of insufficient numbers. We feel that there is enough unintentional bias resulting from working with isolated teeth (see Turnbull 1960 for a discussion of isolated didelphid molars). We have followed Clemens' (1966, p. 3) stylar cusp terminology rather than that of von Koenigswald (1970, p. 8). Abbreviations used are: LACM, Los Angeles County Museum of Natural History; UC, Field Museum of Natural History (University of Chicago collection); SDSM, Museum of Geology, South Dakota School of Mines and Technology; BBQII, Black Bear Quarry II. Specimen numbers are SDSM unless otherwise indicated.

Systematics

Class Mammalia
Order Marsupialia Illiger, 1821
Family Didelphidae Gray, 1821

Peratherium Aymard, 1850

1873. *Herpetotherium* Cope, Palaeont. Bull. No. 16

Peratherium fugax (Cope)
Figs. 1, 2

1873. *Herpetotherium fugax*, Cope
1884. *Peratherium fugax*, Cope
1953. *Peratherium fugax*, Galbreath
1961. *Herpetotherium fugax*, Hough
Galbreath (1953) refers specimens from the Horsetail Creek member (Chadronian) and the Cedar Creek member (Orellan) of the White River Formation of northeastern Colorado to *P. fugax*.

Middle Oligocene Sample—Scenic Member, Brule Formation

LOWER DENTITION
Referred specimens—31135 (Fig. 1a, b) a right mandibular fragment with M_{3-4} from Shannon County, South Dakota (original data given as; Cuny Table, Lower Orellan); and 6413 (Fig. 1c, d), a left mandibular fragment with roots of M_2, and M_{3-4} from Zone D, Insectivore level, Slim Buttes, Harding County, South Dakota (see Lillegraven, 1970). A description of these teeth is essentially that given for the genus by Simpson (1968). The two specimens differ from each other in M_4. In 31135 a distinct cuspid rises from the hypoconid crest in the floor of the talonid basin. The second specimen does not have this cuspid. The difference is interpreted as one of individual variation and not as diagnostic. M_3 in 6413 has a continuous buccal cingulum, though weak, in its passage from trigonid to talonid. Other than these characters the Slim Buttes specimens do not differ from those coming from the Big Badlands. Several other specimens are known from Zone D plus a few from Zone B and Zone E.

UPPER DENTITION
The type of *P. fugax* is a lower dentition. However, upper dentitions have been referred to it, usually because of some proximity (stratigraphic) to lower dentitions. Even Cope (1884) described and referred upper dentitions to *P. fugax*.
 The upper teeth appear to be more variable than the lower. McGrew's (1937) diagnosis of *P. youngi* from the Harrison Formation in Sioux County, Nebraska, separates it from *P. fugax* on stylar cusp differences. The absence of stylar cusp A in *P. youngi* is probably the result of wear. Stylar cusp C is quite large and cusp D is absent. In some of the teeth described below, stylar cusp C is not overly large nor is cusp D. However, with wear the two sometimes fuse into a single large stylar cusp. We believe this is what has happened in the type of *P youngi*. Another char-

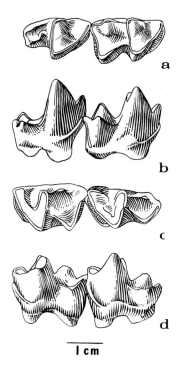

Fig. 1 *Peratherium fugax*, × 10.
   a. SDSM 31135, right M_{3-4}.
   b. SDSM 31135, M_{3-4}, labial view.
   c. SDSM 6413, left M_{3-4}.
   d. SDSM 6413, M_{3-4}, labial view.

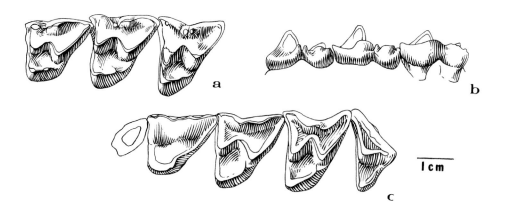

Fig. 2 *Peratherium fugax*, × 10.
   a. SDSM 31134, left M^{1-3}.
   b. SDSM 31134, labial view.
   c. SDSM 63151, left M^{1-4}.

acter McGrew uses to separate *P. youngi* is the position of stylar cusp c in relation to the anterior arm of the metacone. In the type of *P. youngi* it is posterior to
the tip of the anterior arm of the metacone. In the teeth to be described, this is
seen as a variable character.

Referred specimens—Shannon County, South Dakota: 31134 (Fig. 2a, b), a
left maxillary fragment with M^{1-3} has stylar cusps as follows: M^1—A, small; B,
large; C, small, in line with anterior arm of metacone; D, large and worn; E, B, and
D about subequal. M^2—A and B as in M^1; C, small and at base of cingulum a V-
shaped notch, otherwise as in M^1; D, large; E, B, and D subequal. M^3—A, small; B,
a little larger; C and D twinned with C slightly smaller, C in line with anterior arm
of metacone; E, B, and D subequal. Zone B, Slim Buttes, Harding County, South
Dakota: 63151 (Fig. 2c), a left maxillary fragment with M^{1-4}. This specimen is
much more worn than the one from the Big Badlands. In M^{1-3} stylar cusps C and
D are fused. Cusp C is in line with the anterior arm of the metacone in M^{1-2} and
anterior of it in M^3.

Peratherium spindleri Macdonald
Figs. 4a–c; 5a; 6a, b

1963.   *Peratherium spindleri*, Macdonald
1970.   *Peratherium spindleri*, Macdonald

Lower Miocene Sample—Sharps Formation

LOWER DENTITION
Referred specimens—Shannon County, South Dakota: 54343 (type of *P. spind-
leri* Macdonald); 5694, 12 isolated teeth (Godsell Ranch channel fauna); and
LACM 9252. These have been described in detail by Macdonald (1963, 1970).

The diagnostic characters given for *P. spindleri* Macdonald (1963) are: (1)
medium size, (2) strongly developed anterior and posterior cingula on the lower
molars, and (3) labial cingulum continuous. All three of these do distinguish the
type of *P. spindleri*. However, when compared with the referred specimens from
beds of the same age, the size does not deviate from the range normally expected.
The teeth of *Didelphis* vary a great deal in size (Turnbull, 1960). The strongly
developed anterior and posterior cingula of the lower molars is a common feature
in the Oligocene and Miocene specimens. Simpson (1968) does state that the
basal cingulum is anterior only, but this is either in error or a character of Eocene
species of *Peratherium*. The third character of *P. spindleri*, the continuous buccal
cingulum, is well developed in the type but varies from complete to interrupted in
other specimens. This character, seemingly, is highly variable.

The type of *P. spindleri* indicates a smaller than average individual with well-
developed cingula and is in this way distinguished from *P. fugax*. When considered
as part of a population it is not readily separable from other members and scarcely
distinguishable from *P. fugax*.

UPPER DENTITION
Referred specimens—5835 (Figs. 4a, b, c; 5a; 6a, b), nine isolated teeth
(Godsell Ranch channel fauna). These teeth show considerable variation in stylar
cusp C. It is anterior, level, or posterior to the anterior arm of the metacone in M^1.

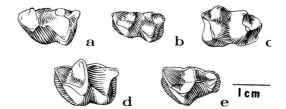

Fig. 3 *Peratherium youngi*, × 10.
   a. SDSM 69155, left M_2.
   Peratherium sp., × 10.
   b. SDSM 67120, left M_1.
   c. SDSM 67126, right M_2.
   d. SDSM 67140, left M_3.
   e. SDSM 67135, left M_4.

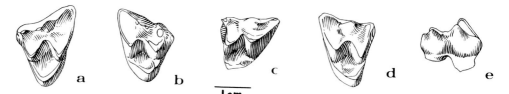

Fig. 4 *Peratherium spindleri*, × 10.
   a. SDSM 5835.15, left M^1.
   b. SDSM 5835.56, right M^1.
   Peratherium sp., × 10.
   c. SDSM 67111, left M^1.
   d. SDSM 67118, right M^1.
   Peratherium spindleri, × 10.
   e. 5835.56, labial view.

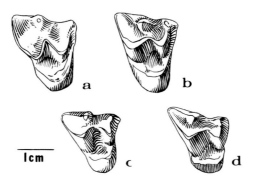

Fig. 5 *Peratherium spindleri*, × 10.
   a. SDSM 5835.19, right M^2.
   Peratherium cf. *spindleri*, × 10.
   b. LACM 23505.37, right M^2.
   Peratherium youngi, × 10.
   c. SDSM 69156, right M^2.
   Peratherium sp., × 10.
   d. SDSM 67109, right M^2.

In one M^2 c and d are fused. In another, c is in line with the anterior arm of the metacone and in a third, may be in advance of it. This last tooth is too worn for positive location of cusp position. Each of three M^3s is different. In one, c is well anterior to the anterior arm of the metacone, in a second it is in line with it, and in the third, c and d are fused.

While the variability in the teeth referred to *P. spindleri* seems to fall within the range of variation seen in *P. fugax*, the size of the sample is too small to allow us to place the name in synonymy.

Peratherium* cf. *spindleri
Figs. 5b; 6c–e

Lower Miocene Sample—Monroe Creek Formation

LOWER DENTITION
Referred specimens—Shannon County, South Dakota: LACM 23505, two M_2s and five M_3s. These do not differ significantly from lower teeth of specimens of Oligocene age.

UPPER DENTITION
Referred specimens—Shannon County, South Dakota: 6292, an M^3 with fused c and d stylar cusps; LACM 23505 (Figs. 5b, 6c, 6d, 6e); (seven isolated teeth), with c and d fused in an M^2, and with c in line with the anterior arm of the metacone in two M^3s and posterior in another in which it is twinned with d.

***Peratherium youngi* McGrew**
Figs. 3a; 5c; 6f

1937. *Peratherium youngi*, McGrew

Lower Miocene Sample—Harrison Formation

LOWER DENTITION
Referred specimen—Shannon County, South Dakota: 69155 (Fig. 3a), a left M_2. The tooth is moderately worn, particularly the entoconid. Part of the paraconid and all of the metaconid are broken off. This tooth generally resembles those from the Monroe Creek.

UPPER DENTITION
Referred specimen—Shannon County, South Dakota: 69156 (Fig. 5c), a right M^2, has a large c cusp which is posterior to the anterior arm of the metacone. 69710, a left M^1, has stylar cusps that are but slightly worn; c is much smaller than d and appressed to it; c is in line with the anterior arm of the metacone, and d posterior to it. 69171, a left M^2, is similar to the preceding. 67157, a left M^3, has subequal c and d cusps that are worn and fused; the c portion of the cusp is in line with the anterior arm of the metacone, and the d portion is posterior to it. 69172, a left M^3, is damaged so that the condition of the c cusp cannot be determined; cusp d is large and posterior to the anterior arm of the metacone.

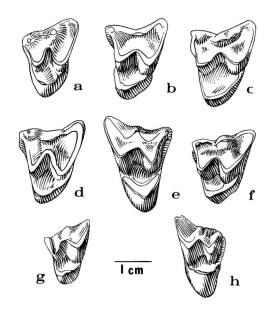

Fig. 6 *Peratherium spindleri,* × 10.   e. LACM 23505.38, right M³.
    a. SDSM 5835.13, left M³.   *Peratherium youngi,* × 10.
    b. SDSM 5835.14, left M³.   f. UC 1544 (type), left M³.
    Peratherium cf. *spindleri,* × 10.   *Peratherium* sp., × 10.
    c. LACM 23505.25, left M³.   g. SDSM 67112, left M³.
    d. LACM 23505.28, left M³.   h. SDSM 67116, right M³.

The type of *P. youngi* McGrew, UC 1544 (Fig. 6f), is from the Harrison Formation in Sioux County, Nebraska. This species, based on a single upper molar (M³) is, so far as we know, the only North American species of the genus based on the upper dentition. Stylar cusp C is large and posterior to the anterior arm of the metacone. Examination of the external wall of the tooth shows that the base of cusp C appears doubled and quite likely the large size is the result of twinning with cusp D.

L.J. Bryant (1969) referred the teeth from the Monroe Creek Formation to *P. youngi*. In addition to the type specimen, the specimens from Shannon County are the only other teeth of the species known from the Harrison. A population sample of five isolated teeth makes verification of the species doubtful.

Peratherium sp.
Figs. 3b–e; 5d; 6g, h

Middle Miocene Sample—Rosebud Formation

LOWER DENTITION
Referred specimens—Bennett County, South Dakota (Black Bear Quarry II):

67119–67137, and 67140 (Figs. 3b, 3c, 3d, 3e). Except for the generally slightly smaller size these teeth are not different from those from the older beds.

UPPER DENTITION

Referred specimens—(as above). 67117 (Fig. 4c), 67108, and 67118 (Fig. 4d, M^1); 67109 (Fig. 5d) and 67113 (M^2); 6796, 67112 (Fig. 6g), 67115, and 67116 (Fig. 6h, M^3).

Lack of sufficient specimens plus the variability noted in teeth of *Peratherium* prevent reliable reference of the teeth from BBQII to any particular species.

Discussion

Simpson (1968) points out that lower molars in *Peratherium* differ little except in size, but that upper molars may be more distinctive among species. Reliance has been placed on the arrangement of cingular cusps in various didelphid genera on the assumption that position and development are consistent from species to species. This may well be true for late Cretaceous species (Clemens, 1966), but within the Oligocene and Miocene species of *Peratherium* we find some variability in the position and development of cusp C.

Of ten M^{1}s examined, two have stylar cusp C anterior to the anterior arm of the metacone, four in line with it, and four posterior to it. In one Orellan specimen C and D are fused. Of 11 M^{2}s, four have stylar cusp C anterior to the anterior arm of the metacone, four in line with it, and three posterior to it. In one Orellan specimen C and D are fused as a result of wear. All three teeth from the Sharps Formation have fused C and D cusps as do both teeth from the Monroe Creek. One of the Harrison M^{2}s has fused cusps. Of 16 M^{3}s, two have stylar cusp C anterior to the anterior arm of the metacone, ten are in line with it, and four are posterior. C and D cusps are fused in both Orellan specimens, one Sharps, one Monroe Creek, and one Harrison specimen.

Variation in position of stylar cusp C, arranged by formation or member, is given in Table 2.

We believe that even this small sample shows the variability of stylar cusp C's position, although there might have been a general shift to a more posterad position of the cusp in later populations. It would be difficult to draw a definite line because the samples may represent a cline of variation. The samples, however, are too small for certain determination. With more specimens it might be possible to show that cusp C's position is variable in all populations.

Comparison of teeth from Black Bear Quarry II with specimens referred to *P. fugax* (Cope) from the Oligocene indicated that a slight but definite size difference was observable. It appeared that a possible separate smaller species might be distinguished from the Rosebud Formation. Comparison of the Rosebud material with *P. spindleri* Macdonald and referred specimens (all from the early Arikareean Sharps Formation) did not seem to make so clear-cut a distinction. Additional comparisons were made with specimens of *Peratherium* from the Monroe Creek Formation (South Dakota) and with all known specimens from the Harrison Formation. There appears to be a slight decrease in size in specimens from Oligocene through to the early Middle Miocene.

Table 1. Measurements of teeth of *Peratherium* from various formations. A-P = antero-posterior and Tr = transverse diameter.

M₁	A-P	Tr
Sharps Formation 54343	1.42	0.8
5835.20	1.65	0.9
5835.37	1.65	0.85
Rosebud Formation 67119	1.8	0.8
67120	1.4	0.8
67121	1.85	0.9
67122	1.65	0.8
67123	1.55	0.8
67124	1.55	0.8
67125	1.6	0.8
M₂		
Orella member Big Badlands 31136	1.8	1.15
Orella member Slim Buttes 62117	1.9	1.1
Sharps Formation 54343	1.65	0.92
5835.1	1.7	0.9
5835.31	1.7	1.0
LACM 9252	1.65	1.07
Monroe Creek Formation LACM 23505.3	1.79	1.0
LACM 23505.5	1.75	1.5
Harrison Formation 69155	1.7	1.0
Rosebud Formation 67126	1.7	1.05
67127	1.55	1.0
67128	1.49	0.8
67130	1.6	0.85
67131	1.6	0.75
M₃		
Orella member Big Badlands 31135	2.0	1.1
31136	1.95	1.25
Orella member Slim Buttes 62117	2.0	1.1
6413	2.0	1.2
63683	1.8	1.1
Sharps Formation LACM 9252	1.75	1.15
54343	1.7	1.0
5835.2	1.9	1.09
5835.18	2.0	1.19
5835.32	1.8	1.0
Monroe Creek Formation LACM 23505.1	1.89	1.5
LACM 23505.4	1.9	1.5
LACM 23505.6	1.9	1.3
LACM 23505.16	1.85	1.15
LACM 23505.14	1.85	1.1
Rosebud Formation 67132	1.85	0.8
67133	1.8	1.2
67134	1.6	1.1
67140	1.71	0.91
M₄		
Orella member Big Badlands 31135	1.9	0.95
31136	2.1	1.0

Table 1, concluded.

Orella member Slim Buttes 6413	2.0	1.1
63683	1.8	0.95
63102	2.05	0.9
65510	1.8	1.05
Sharps Formation LACM 9252	1.7	0.99
54343	1.62	0.9
5694	1.8	1.0
Rosebud Formation 67135	1.6	0.99
67136	1.75	0.9
67137	1.5	1.0

M^1

Orella member Big Badlands 31134	1.85	1.55
Orella member Slim Buttes 63151	1.85	1.8
Sharps Formation 5835.15	1.79	1.79
5835.49	1.9	1.8
5835.56	1.8	1.6
Harrison Formation 69170	1.8	1.56
Rosebud Formation 67117	1.76	1.41
67108	1.75	1.8
67111	1.75	1.4

M^2

Orella member Big Badlands 31134	1.75	1.85
Orella member Slim Buttes 63151	1.75	1.9
Sharps Formation 5835.4	1.85	1.85
5835.19	1.79	1.85
5835.57	1.8	1.9
Monroe Creek Formation LACM 23505.37	1.6	2.0
LACM 23505.39	1.9	2.0
Harrison Formation 69156	1.58	1.5
69171	1.81	1.45
Rosebud Formation 67109	1.65	1.5
67113	broken	1.65

M^3

Orella member Big Badlands 31134	1.72	2.0
Orella member Slim Buttes 63151	1.75	2.15
Sharps Formation 5835.13	1.59	1.75
5835.14	1.69	2.09
5835.59	1.68	1.95
Monroe Creek Formation LACM 23505.25	1.9	2.19
LACM 23505.28	1.9	2.3
LACM 23505.38	1.75	2.2
6292	1.68	1.95
Harrison Formation UC 1544	1.85	1.65
67157	1.72	2.05
69172	1.82	2.19
Rosebud Formation 67112	1.41	1.4
67115	1.6	1.79

Table 2. Position of stylar cusp C relative to the anterior arm of the metacone.

Tooth and formation or member	Anterior	In line	Posterior
M¹			
Orella		2	
Sharps	1	1	1
Harrison	1		
Rosebud		1	3
M²			
Orella		2	
Sharps	1	1	1
Monroe Creek	1	1	
Harrison	1		1
Rosebud	1		1
M³			
Orella		2	
Sharps	1	2	
Monroe Creek	1	3	
Harrison		2	1 (type)
Rosebud		1	3

Conclusions

The results of this study present a taxonomic dilemma. The Oligocene species have been called *P. fugax*; the Sharps Formation species, *P. spindleri*; the Monroe Creek species, either *P. spindleri* or *P. youngi*; and the Harrison species, *P. youngi*. This is no more than assigning a different name for each stratigraphic unit. Possibly these names represent populations of the same species changing but slightly through time. There are a number of solutions to such situations. One would be to leave the names as they are, that is, stratigraphic species. A second is also the possibility that there are two co-existent species. A third is to consider them all as one species with temporal as well as geographic populations. While our inclination is toward this last approach, we do not feel that we have sufficient evidence to justify it. We have tried to decide between taxonomic reality and taxonomic practicality. It is a question that must be decided whenever we are confronted with chronoclines.

The implications in this paper are that: (1) although several species of the genus *Peratherium* from the Middle Tertiary of the Great Plains have been described, the evidence upon which they have been based is meager (e.g. Galbreath (1953) has inferred that *P. hunti* (Cope) is conspecific with *Nanodelphys minutus* McGrew, and there seems no doubt that this inference is correct. This paper appears to have been overlooked by Hough (1961)); and (2) not enough additional material is available to test the validity of these taxa; if we assume that the specimens discussed in this paper represent samples of a continuous evolutionary line, then there was a decrease in average size beginning with the *P. fugax* population in the early Oligocene through to the population from Black Bear Quarry II.

The discovery of *Peratherium* in the Black Bear Quarry II local fauna extends its temporal range from the Harrison Formation into the early Middle Miocene in North America. To date, only isolated lower and upper molars are known from BBQII. These additions give us a continuous record of *Peratherium* from middle Chadronian through the early Hemingfordian in a limited geographic area.

It seems proper, at this time, to provide a provisional faunal list of the Black Bear Quarry II local fauna for use by other workers:

Marsupialia
  Didelphidae
    Peratherium sp.

Insectivora
  Erinaceidae
    Amphechinus sp.
  Soricidae
    Heterosorex sp.
  Talpidae
    Scalopoides sp.
    Mesoscalops sp.
    Genera not determined

Lagomorpha
  Leporidae
    Archaeolagus macrocephalus
    Archaeolagus primigenius
  Ochotonidae
    Gripholagomys lavocati

Rodentia
  Eomyidae
    Genus not determined
  Sciuridae
    Genera not determined
  Cricetidae
    Genera not determined
  Geomyidae
    Dikkomys sp.
    Genera not determined
  Heteromyidae
    Genera not determined
  Zapodidae
    Plesiosminthus sp.

Acknowledgments

We are indebted to William D. Turnbull of the Field Museum of Natural History and to David P. Whistler of the Natural History Museum, Los Angeles County for the loan of specimens. The drawings were made by Merton C. Bowman. The presence of a didelphid in the collection was called to our attention by Malcolm C. McKenna during a cursory examination of BBQII specimens. Special appreciation for criticism of preliminary manuscripts is extended to P.R. Bjork, R.H. Tedford, and R.W. Wilson.

Literature Cited

ALLEN, J.A.
  1900   Descriptions of new American marsupials. Bull. Amer. Mus. Nat. Hist., 13: 191–199.

BRYANT, L.J.
  1969   MS thesis, South Dakota School of Mines and Technology. Unpublished.

BUMP, J.D.
1951   Guide Book, Fifth Field Conference of the Society of Vertebrate Paleontology in Western South Dakota. Rapid City. 85 pp.

CLARK, J.
1937   The stratigraphy and paleontology of the Chadron Formation in the Big Badlands of South Dakota. Ann. Carnegie Mus., 25: 261–350.

CLEMENS, W.A., JR.
1966   Fossil mammals of the type Lance Formation, Wyoming. Univ. Calif. Publ. Geol. Sci., 62: 1–122.

COPE, E.D.
1884   The Vertebrata of the Tertiary formations of the West. Report U.S. Geol. Surv. Terr., F.V. Hayden in charge. Washington. 1009 pp.

GALBREATH, E.C.
1953   A contribution to the Tertiary geology and paleontology of northeastern Colorado. Univ. Kansas Paleo. Contrib., 13(art. 4): 1–120.

GREEN, M.
1972   Lagomorpha from the Rosebud Formation of South Dakota. Jour. Paleontology, 46: 377–385.

HOUGH, J.R.
1961   Review of Oligocene didelphid marsupials. Jour. Paleontology, 35: 218–228.

KOENIGSWALD, W. VON
1970   *Peratherium* (Marsupialia) im Ober-Oligozän und Miozän von Europa. Bayer. Akad. Wissen. Math.-Nat. Kl., Abh., n.f., 144: 1–79.

LAVOCAT, R.
1952   Révision de la faune des mammifères Oligocènes d'Auvergne et du Velay. Paris. 153 pp.

MACDONALD, J.R.
1963   The Miocene faunas from the Wounded Knee area of western South Dakota. Bull. Amer. Mus. Nat. Hist., 125: 139–238.
1970   Review of the Miocene Wounded Knee faunas of southwestern South Dakota. Bull. Los Angeles County Mus. Nat. Hist., 8: 1–82.

MCGREW, P.O.
1937   New marsupials from the Tertiary of Nebraska. Jour. Geol., 45: 448–455.

SCOTT, W.B. AND G.L. JEPSEN
1941   The mammalian fauna of the White River Oligocene. Part v. Trans. Amer. Philos. Soc., 28: 747–980.

SIMPSON, G.G.
1968   A didelphid (Marsupialia) from the early Eocene of Colorado. Postilla, 15: 1–3.

TATE, G.H.H.
1933   A systematic revision of the marsupial genus, *Marmosa*, with a discussion of the adaptive radiation of the murine oppossums (*Marmosa*). Bull. Amer. Mus. Nat. Hist., 66: 1–250.

TURNBULL, W.D.
1960   A Lance didelphid molar, with comments on the problem of Lance therians. Fieldiana—Geol., 10: 525–537.

WILSON, R.W.
1960   Early Miocene rodents and insectivores from northeastern Colorado. Univ. Kansas Paleo. Contrib., 24(art. 7): 1–92.

Restoration of Masticatory Musculature of *Thylacosmilus*

William D. Turnbull
Curator, Fossil Mammals, Field Museum of
Natural History, Chicago, and Research Associate, Texas
Memorial Museum, Austin

Abstract

The jaw muscles of the remarkably specialized, extinct sabre-tooth marsupial *Thylacosmilus* are restored. The restorations are based upon details of the origin and insertion scars and spatial relationships of each muscle, all of which are described. These constraints considerably limit the possibilities and lend credibility to the restoration. Relative proportions of the jaw muscles are determined. These compare well with those of both carnivores and generalized mammals.

Introduction

Comparative studies of the jaw musculature of Recent mammals provide the basis for interpretation of their counterparts in fossil mammals (Adams, 1919; Becht, 1953; Maynard Smith and Savage, 1959; Schumacher, 1961; Turnbull, 1970; Finch, MS). This is especially true for fossil taxa that are generalized in their masticatory apparatus (i.e., have essentially the ancestral, primitive pattern), or for those closely related to a living specialized form.

Thylacosmilus, which was first described by one of my predecessors at the Field Museum (Riggs, 1934), does not fit into either of these categories for it belongs to a terminal branch of a highly specialized marsupial carnivore stock and has no close living relatives. Yet the fact that its specialization is in a carnivorous direction means that its masticatory musculature can be expected to have been fairly generalized as is the case with the living representatives of the group designated (Turnbull, 1970) as Specialized Group I (Carnivore-Shear Group). Of my three specialized groups, this one includes forms that may have highly modified dentitions, but invariably this apparatus is operated by a musculature that is only slightly altered from that of the Generalized Group. Thus, the dental specializations do not require major modifications of the musculature to permit their function. This, and the fact that most of the scars of the jaw muscles do not significantly depart from the standard patterns of either the Generalized Group or Specialized Group I (Carnivore-Shear Group), led me to conclude that I could make a restoration with reasonable confidence.

Thylacosmilus specialized its dentition in a remarkable fashion – its long sabre-like upper canine teeth are hyperdeveloped and arced in a manner that is strikingly parallel to, and even considerably exceeds in degree, that of the advanced sabre-toothed cats (e.g., *Smilodon, Dinobastis*). The rest of the dentition also shows carnivorous specialization, but in this it is not outwardly so obvious. The premolars and molars are reduced to narrow linear blades that functioned by orthal jaw closure and were capable of slicing flesh into pieces small enough to be swallowed. They appear to be far too delicately built to have been able to perform crushing, bone-breaking, or grinding functions.

Directly related to the hyperdeveloped upper canines is an equally hyperdeveloped cervical musculature. This is reflected in the expansion and nature of the surfaces of attachment on the occipital and basicranial portions of the skull, where the insertion occurs, and on the centrum and various processes of each of the anterior five cervical vertebrae where they originate. Riggs (1934, pp. 11–12, 22–25) noted these details for each region and his figures (Pls. I–VI) adequately demonstrate the conditions that pertain. He makes the following summary statement (pp. 20–21):

The extreme specialization observed in the skull and mandible of *Thylacosmilus* has evidently resulted from a series of adaptations in animals of this line. The moving factor in these adaptations has been the extraordinary development of the upper canine teeth, accompanied, no doubt, by highly specialized preying habits acquired by the animal. Great stress was being laid upon the canine. As a corollary of the development of these weapons, there are observed evidences of great relative strength in the vertebrae of the neck and in those muscular attachments of the head which give insertion to muscles at the occiput, at the basicranium, and at the ventral surface of the centra of the cervical vertebrae. While all these attachments give evidence of great strength, the series at the ventral surface of the cranium and at the ventral processes of the cervical vertebrae, which give insertion to muscles used in forcing the head downward, have an unusual development. This fact strongly indicates that they must have been used in a downward stroke of the saberteeth, or in retaining a hold secured by those weapons.

If there is a fault to be found with Riggs' statement, in my opinion, it is that he is perhaps too conservative. The protruberances, rugosities and excrescences in the areas of cranial attachment of the cervical musculature are so pronounced, and the fields themselves so expanded (especially relative to size of the braincase), that one cannot overemphasize the implication for the strength of the musculature.

In *Thylacosmilus*, as in the advanced sabre-tooth cats, there is hyperdevelopment of the mastoid process (Matthew, 1910, p. 296). This area provides the insertion of the cleidomastoid and sternomastoid muscles. Such development shows that these muscles must have been exceptionally powerful and hence capable of causing the head and neck to flex with great force.

The basilar (basisphenoid) tubercles surely indicate a massive tendinous insertion of the M. rectus capitis ventralis major (= M. longus capitis of authors including Riggs), and the area immediately behind the tubercles must have served for insertion of the M. rectus capitis ventralis minor ($=$ M. rectus capitis ventralis of Riggs and others). In detail the latter field appears to have been complex, with a tendinous lateral portion that was contiguous with the tendon of the former muscle, and which had both tendinous and fleshy attachment surfaces posteromedially.

Every feature of these combined fields for the insertion of the musculature that produces flexion of the head is remarkable, more remarkable even than those for the very obvious dorsal cervical musculature, which in quadrupedal mammals is usually well developed in order to support the cantilevered head.

I believe that this head-controlling musculature must have been not only extremely powerful, but also superbly capable of manoeuvring the head and its attached canine blades with great precision. Otherwise there would be risk of an angled or transverse blow breaking the greatly elongated and unprotected canines. Their firm anchorage to the skull, and their curved, dagger-like shape make the canines well suited for both powerful stabbing, and slashing and slicing (see discussion by Simpson, 1941). The stabbing would be accomplished by the momentum of a lunge and by the downward flexion of head and neck, while the slicing and slashing would result from additional flexion and retraction of the head combined with a pulling back of the head once the canines have penetrated flesh. I envisage this stabbing, slashing and slicing sequence to have resulted from one integrated continuous set of movements, with the slashing-slicing feature being equal to the stabbing one in its degree of refinement. That the canines themselves must have been quite vulnerable to damage from a forceful lateral blow is indicated by their cross-section which is thin mediolaterally and by the presence of the full-length symphyseal "scabbard" or "sheath".

Methods and Materials

The procedures followed for restoring the jaw muscles are given below:

1. The restored skull and jaws are cast and then mounted with the jaws slightly agape.

2. The attachment fields of each of the jaw muscles are located and designated, including, when apparent, subdivisions of the major muscles, based upon details of the muscle scars themselves and the known regions of the origin and insertions described for Recent mammals.

3. Using Plastilina, each muscle in turn, beginning with the deepest, is modelled so that the spatial relations and attachments are observed and each muscle appears to be functionally workable.

4. After completion of step 3, the model is removed, a two-piece plaster mould is made, and the "muscle" is then cast in a flexible silicone or polysulphide rubber product. (I used tan Smooth-On FMC #100 from Smooth-On Products, Inc., 1000 Valley Road, Gillette, New Jersey, U.S.A. 07933.)

5. This rubber "muscle" is then lightly glued into position on the plaster replica of the skull, and the next-most superficial muscle modelled, cast and glued into

position, repeating steps 3, 4 and 5 for this and each succeeding muscle until all have been restored.

6. At each succeeding stage, the reasonableness, not only of the muscle being restored, but of the cumulative set of muscle masses and their spatial array, affords a fresh opportunity to assess whether or not a functional integrity has been achieved.

7. Finally each restored rubber muscle is removed and weighed to calculate its percentage of the total jaw musculature. In this way direct comparisons with percentage weights of muscles from many Recent mammals become possible (Turnbull, 1970, Tables x–xii, and figs. 34–37) and provide the basis for a first estimate of forces and calculations of useful power (Turnbull, 1970, pp. 278–296).

The key to the success of this method lies in the repeated opportunities for reassessment, each with the constraints imposed by the attachment areas and the other three-dimensional spatial relationships. The method constitutes a slight refinement over that used by Finch (pers. comm.) for *Thylacoleo* in which Plastilina "muscles" were modelled and kept distinct by thin clear plastic separators instead of using rubber "muscles". Either way, any gross error in the restoration of one muscle so permeates the whole that it becomes difficult to achieve a reasonable restoration for the succeeding layer. Thus repeated checks are present. In *Thylacosmilus* only one area of attachment proved to be troublesome, that of the posterior digastric (see below). While this has not seriously affected the final result, it would have been cause for greater concern had it not been a minor part of the total jaw musculature.

The materials used constitute the entire sample of *Thylacosmilus* in the collection of the Field Museum of Natural History:

T. atrox Riggs, P 14531, the holotype, consisting of a skull and parts of the skeleton;

T. atrox Riggs, P 14344, paratype, a cranium, mandible, and parts of the skeleton; and

T. lentis Riggs, P 14474, holotype, an incomplete skull with dentition.

All are from Catamarca, Argentina. The two holotypes are from the Middle Pliocene, Corral Quemada Formation at Corral Quemada, Departmente Belén, Provincia de Catamarca and the second *T. atrox*, specimen P 14344, is from a slightly lower level of lower Middle Pliocene Age, Araucano Formation, Chiquimil, Rio Santa Maria.

Description

Muscle Attachment Areas

M. pterygoideus externus. The origin appears to have been from the posterior half of the wing of the pterygoid bone, from an elongate, ovate, ventral-facing area situated anterolateral to the prominent protruberances of the basisphenoid (Figs. 1B, 2B). The attachment was probably both tendinous and fleshy, the tendinous portion enveloping the fleshy part. The insertion, as nearly as can be ascertained from *P* 14344, was on the usual area, the medial edge of the condyle, and probably it had a tapering wedge extending anteriorly along the neck of the condyle as in many living mammals (Fig. 1B). The restored muscle, seen in ventral view (Fig. 4A), has a flattened conical form (but see further the discussion of the next muscle and Fig. 4D).

M. pterygoideus internus. This muscle takes origin at the ventrolateral edge of the pterygoid bone, from a continuous and twisted surface which faces more ventrally in its anterior portion, just behind the palate, and becomes directed laterally in its posterior part near the optic foramen (Figs. 1B, 2B). There appears to have been another head which originated from the ? alisphenoid behind the optic foramen, beneath the deepest part of the M. temporalis, pars profunda, but this head could also have been a part of the preceding muscle. If that were the case, the proportions of the two muscles would clearly be significantly altered, but the total of the pterygoid complex would remain unchanged. My decision to consider this surface to be for the origin of a second head of the internal pterygoid rests upon the closer similarity to the condition of the muscles in the living *Didelphys virginiana* than does the alternative restoration. The anterior part of the origin appears to have been mostly fleshy with a tendinous envelopment, especially medially. Insertion was entirely by a heavy tendinous band to judge by the rugose scar on the inflected angular process. The field of insertion is elongate, expanded posteriorly and tapered anteriorly (Fig. 1B). The inferior dental foramen lies just above the middle of the insertion, between it and that of the deep temporalis, much as in *Didelphis*.

M. temporalis, pars profunda. This muscle was thick, filling much of the temporal fossa posterolaterally but appreciably less anteriorly (Figs. 1A, 2A). It clearly contained very extensive tendinous strands and laminae, as the area of origin is extremely rough with many irregular bony prominences. It is not distinctly separated from the superficial temporalis except anteriorly. Probably its fibres mixed with those of the superficial temporalis over much of their continuous areas of origin. This is frequently the case in living mammals and, when so, it is usually possible to force a separation by working from the insertions where there is never any fusion. Insertion of the deep temporalis was as usual on the planum tendineum temporalis, most of the medial aspect of the coronoid process and part of the body of the ramus (Fig. 1A). There is no clear-cut indication of the limits of the insertion on the fossils, but neither is there any reason to believe that it was in any way unusual. I interpret this smooth field to imply that the insertion was entirely fleshy. The chief problem with this muscle is to determine just how much of it inserted onto the planum tendineum temporalis, as is also the case with the superficial temporalis. The planum appears to have been tremendous, to judge by the scar on the coronoid process, where all of the muscle's tip and approximately the upper half of the medial side and the anterolateral facet on the lateral side inserted. From this I assume that a significant portion of both muscles had their fibres inserted into the planum, as in many living mammals.

M. temporalis, pars superficialis. The superficial temporalis, like the deep temporalis, was a large, thick muscle mass. It originated from the entire area of the temporal fossa not already occupied by the deep portion: from the posterolaterally facing area of the front of the fossa and from the raised parts of both lambdoidal and sagittal crests (Figs. 1A, 2A). Since the temporalis is in most mammals nearly always covered by a heavy aponeurosis and since the superficial edge of the fossa so indicates, I am certain that this condition prevailed in *Thylacosmilus* and that there was an exceptionally thick sheet. This implies that the origin of the muscle extended from the medial side of the aponeurosis for at least its dorsal two-thirds, as in those living forms with a heavy aponeurosis. Insertion was onto the lateral face of the coronoid process to a triangular field (Fig. 1B) and presumably onto the superficial surface of the planum tendineum temporalis.

I assume that there was a pars zygomatica subdivision of the pars superficialis because the zygomatic arch is stout and relatively high, and because one is usually developed when such conditions pertain. In neither of the two specimens in which any part of the arch is present is the bone surface well enough preserved for me to be certain about the limits of this muscle mass, and I have somewhat arbitrarily followed the squamosal-jugal suture as its ventral limit (Figs. 1A, 2A, 2B; A reinforcing bar of plaster partially obscures the area in P 14531, the most complete specimen). This slip is never a very considerable portion of the muscle nor is it usually clearly distinct in its insertion. Rather, it is usually fused with the rest of the superficial part, so I have not worried about the uncertainty.

M. zygomaticomandibularis. The field of origin (Figs. 1A, 1B) as I have interpreted it is somewhat triangular and extends over the entire ventral half of the medial side of the zygomatic arch. Also, in its anterior half, in the area of the union of the postorbital portions of the frontal bones, it overlaps the suture between jugal and the postorbital bar. As noted above, the area is largely covered, so I cannot determine how tendinous or fleshy the origin might have been. Insertion of this muscle was into a recessed oval area deep within the upper part of the masseteric fossa (Fig. 1A). It appears to have been fleshy but with thin tendinous bands throughout, especially in its anterior half, to judge from the complete mandible P 14344.

M. masseter, pars profunda. The deep masseter originates from a well-defined elongated area that occupies approximately the ventral third of the lateral surface of the zygomatic arch, and which extends forward onto the anterior buttress of the arch above the last molar tooth (Fig. 1B, 2B). The area is slightly indented and the rugosities in its anterior portion, which is thicker than the rest, indicate that the muscle was more tendinous anteriorly than elsewhere. Probably it was completely encased in a tendous envelope. Insertion is obviously into the well-developed masseteric fossa, superficial to and also extending anterior to that of the M. zygomaticomandibularis. However, in none of the specimens is there a very prominent scar to indicate unequivocally the limits between these two muscles. Specimen

Fig. 1 Drawings of the skull and jaws of *Thylacosmilus* in lateral view with the attachment areas of the jaw muscles mapped in red. Solid red lines indicate limits that could be ascertained with reasonable certainty; dotted red lines show limits that had to be assumed since no muscle scars were evident, and dashed red lines indicate limits that are hidden in this view. For clarity two diagrams are provided. Drawings of skull with jaw from Riggs (1934).

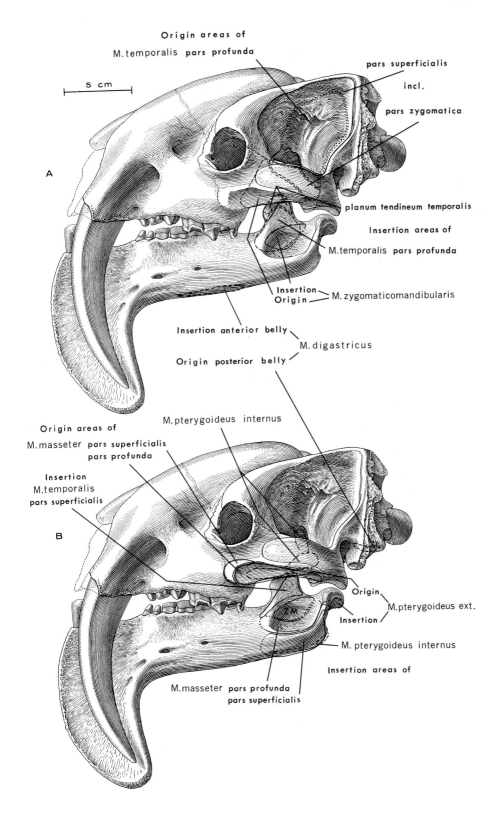

Origin areas of
M.temporalis pars profunda

pars superficialis

incl.

pars zygomatica

5 cm

A

planum tendineum temporalis

Insertion areas of

M.temporalis pars profunda

Insertion
Origin ——— M. zygomaticomandibularis

Insertion anterior belly

M.digastricus

Origin posterior belly

M.pterygoideus internus

Origin areas of
M.masseter pars superficialis
pars profunda

Insertion
M.temporalis
pars superficialis

B

Origin

M.pterygoideus ext.

Insertion

M. pterygoideus internus

Insertion areas of

M.masseter pars profunda
pars superficialis

P 14531 preserves a suggestion of a linear rugosity which must be near the division of the fields of insertion within the fossa, and I have assumed that it actually represents that division and have restored the muscles accordingly (Fig. 1B). The insertion appears to have been mainly fleshy, but with some light tendinous bands in the anterior, and much more tendinous at its posterior edge, where the rim of the fossa is heavily scarred. I have not been able to ascertain confidently the anterior limit of the muscle, and after much thought have finally chosen to restore it as reaching forward to a point just beyond the anterior point of the insertion of M. temporalis, a common condition in living generalized and carnivorous mammals.

M. masseter, pars superficialis. The origin of this muscle was, I believe, almost completely tendinous, as it is in a great many living mammals. The origin is a crescentic prominence, thickest over the anterior end of the origin of the deep masseter. It is contiguous with the enveloping anteroventral tendinous cover of that muscle, as the very sharp and clear attachment scars indicate (Fig. 1B, 2B). Above, in its upper part where the crescent is attenuated (about over the middle of the field of origin of the deep masseter) the scar is indistinct and lacks rugosity. Thus I assume that in this area it was very thin and perhaps, in part, the muscle originated from the superficial tendinous surface of the deep masseter. The insertion was probably quite fleshy, for the most part onto the ventrally- or ventro-laterally-facing triangular flange of the angle of the jaw. No doubt it was tendinous at its posteroventral limits on the angle where it inserted near the insertion of the M. pterygoideus internus, and along the posterior edge of the inflected angle and neck (Fig. 1B). Almost certainly there was a well-developed aponeurosis covering the entire masseteric complex, and probably the superficial masseter was intimately fused with it, to judge by the attachment scars (the usual condition in most other mammals).

M. digastricus. This muscle is assumed to have had the usual two-bellied form with a dividing tendon. In addition, it may have had a raphic tie from the dividing tendon to the ventral neck musculature (see below). The field of insertion of the anterior belly is distinct, but the origin of the posterior belly is uncertain. It probably was from the small posteromedial triangular zone on the exoccipital bone where it laps the mastoid, as restored in Figs. 1B, 2B. There is, as Riggs (1934) noted, no developed paroccipital process, the usual site of origin of this muscle as well as of the Mm. mylohyoideus and stylohyoideus. This lack makes interpretation difficult and far more speculative than for the rest of the musculature. Insertion of the anterior digastric appears to be quite standard—mostly tendinous onto a well-defined, elongate, tapered, rugose area, about one-third as long as that portion of the horizontal ramus which lies behind the symphyseal flange (Fig. 1A). Almost certainly there was a fanning of its superficial tendinous sheet ventromedially between and beneath the rami. Such a condition prevails in a wide variety of living mammals. Probably there was either a tendinous inscription of sorts, or a fibrous mixing of the medial fibres with those of its contralateral member.

Origin areas of

M. temporalis **pars superficialis**

pars zygomatica

pars profunda

attachment of
temporal apponeurosis

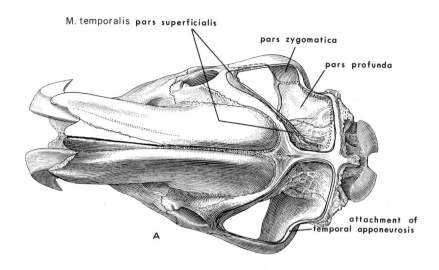

A

Origin areas of

5 cm

M. temporalis **pars zygomatica**

pars profunda

pars superficialis

M. pterygoideus internus

externus

M. zygomaticomandibularis

M. digastricus
posterior belly

B

M. masseter **pars superficialis**

pars profunda

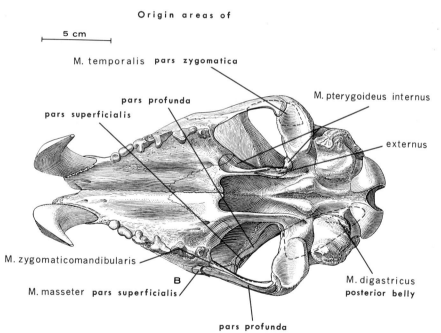

Fig. 2 Drawings of the skull and jaws of *Thylacosmilus* in ventral view showing muscle attachment fields as in Fig. 1.

Problem of the Digastric

The usual arrangement of the digastric muscles of living mammals is such that they come close to being aligned with the axes of the mandibular bodies, and thus they have only a small angulation to their force vectors. This means that at best they cannot be very effective at jaw opening because such small leverage is inefficient. Apparently the demands are normally not great, and with the help of gravity jaw opening is not a problem.

For *Thylacosmilus* the situation was different, because the demands would have been increased while the alignment of the muscle appears to have been such as to make jaw opening less efficient. With its anterior digastric inserted in the usual position and the posterior belly coming directly from the skull base instead of from a protruding paroccipital process, the angular vector is virtually eliminated. Thus it seems unlikely that the usual arrangement of the digastric muscles could by itself have been capable of causing the extreme jaw opening necessary to allow the canines to function. Simpson (1941, p. 9), in his analysis of the function of sabre-teeth, discussed earlier notions about the ability (or lack of ability) to achieve an adequate gape, accepted Matthew's (1910) and Bohlin's (1940, p. 164 ff.) views as regards *Smilodon*, and concluded that of course they did achieve the needed gape. However, he made no mention of the musculature involved.

Since in *Thylacosmilus* there is no direct indication of any other sort of digastric development that could compensate for the missing paroccipital processes to provide a means of implementing extreme jaw opening, I hypothesize that there may have been a tie to the powerful ventral neck musculature, probably by a raphic band connecting to the dividing tendon. Possibly it was also tied to a tendinous envelope of the posterior belly itself. With such a mechanism, flexure of head and neck would automatically assist in initiating and maintaining jaw opening by the resulting backward and downward pull on the dividing tendon. There would result a neat and effective means of assuring that the jaws would be out of the way during thrusting, slashing and slicing strokes of the head.

Muscle Restorations

M. masseter complex was doubtless covered by a tendinous aponeurosis. The pars superficialis as restored has a high trapezoidal form in side view (Fig. 3A, 4A). It appears to have had an entirely tendinous origin and a thick fleshy mass beneath a thinning tendinous cover in its ventral two-thirds. Average fibre direction seems to have been at about 45° to the horizontal.

The pars profunda has a more nearly square form than the superficial portion (Figs. 3B, 4B) and was mostly covered by it; only a wedge-shaped area, its posterodorsal third, was not covered. It was probably more fleshy, but with tendinous strands at various depths within itself. In side-view its fibres appear to have been inclined at about 60°–70° to the horizontal, appreciably more nearly vertical than that of the superficial part.

The M. zygomaticomandibularis also appears to be rather square in outline, but, because most of it is laterally concealed by the arch, it actually has a rectangular form (Fig. 3C). It probably had a solid fleshy mass with many tendinous bands throughout. Its average fibre direction must have been at about 80° to the

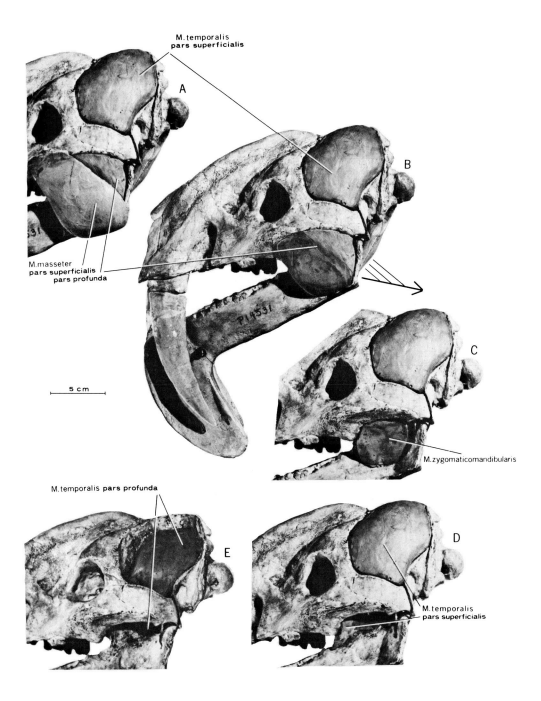

M.temporalis
pars superficialis

A

B

M.masseter
pars superficialis
pars profunda

5 cm

C

M.zygomaticomandibularis

M.temporalis pars profunda

E

D

M.temporalis
pars superficialis

Fig. 3 Photos of the casts of skull and jaws of *Thylacosmilus* with the restored jaw musculature shown in lateral view.

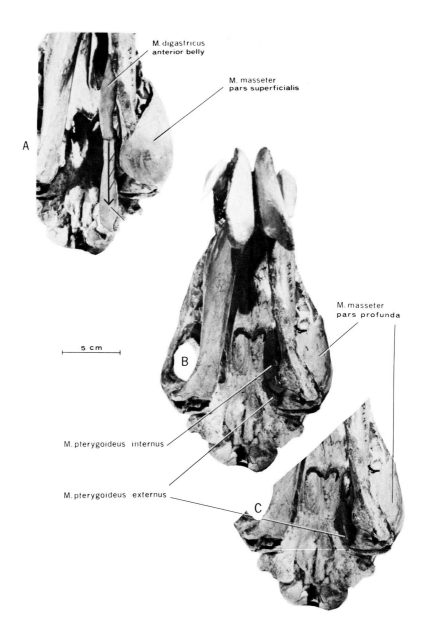

Fig. 4 Photos of the casts of skull and jaws of *Thylacosmilus* with the restored jaw musculature shown in ventral view.

horizontal in side-view, and probably was not too far from this in the transverse plane as well.

M. temporalis complex was overlain by a very heavy temporal aponeurosis (Fig. 2A), and doubtless all of the most superficial laminae (and in echelon, succeedingly deeper ones) of the pars superficialis took origin from it. The pars superficialis (Fig. 3B,D) was thick and arched in the transverse plane. In lateral view its fibres were somewhat converging, inclined at about 110° to the horizontal. I have assumed that there was a small pars zygomatica (Figs. 2A,B) that probably was not distinctly delimited from either the superficial or deep portions of the muscle. Its fibres would have converged somewhat toward their insertions, seen in side-view, and, on the average, they would have been more nearly vertical than those of the other divisions.

The pars profunda (Fig. 3E) must have had many tendinous bands throughout. It was thick and elongated and, in side view, its fibre direction must have been much as in the pars superficialis.

M. pterygoideus complex appears to have been quite standard with the M. pterygoideus externus (Figs. 4B,C), being small and obliquely inclined at about 45°–50° in both planes. The M. pterygoideus internus (Fig. 4B) appears to have been a solid block of tissue whose fibres pulled nearly vertically at about 80° to the horizontal. By working synergistically with the superficial masseter, it would aid in closing the jaws. At the same time it should have been capable of exerting considerable antagonistic force to the rotational pressures along the axis of the mandible that would have been caused by contraction of superficial temporalis, zygomaticomandibularis, and both parts of the masseter.

M. digastricus has already been partially discussed above because of the problems it poses. My hypothesis of a raphic tie to the ventral neck musculature assumes that this may have been by a tendinous connecting band or sheet stoutly fused to the dividing tendon (Figs. 3B, 4A, arrows), and possibly also even to the ventral edge of a tendinous envelopment of the posterior digastric. Thus contraction of the ventral neck musculature would not only cause the head and neck to be flexed, but it would pull back on the dividing tendon of the digastric, giving greatly increased angular vectors to both anterior and posterior digastrics relative to the mandibular axis. In this way not only would the digastrics be better oriented for them to function, but also this would happen automatically with every powerful contraction of the flexing muscles of the head and neck.

Muscle Proportions and Useful Power Calculations

Table 1 gives the weights of each restored muscle and gives jaw muscle proportions in the customary two ways: (1) as percentage of the total of the jaw-closing set of one side (i.e., all but the digastrics), and (2) as percentage of all of the jaw muscles of that side. This way of approximating the true proportions involves the assumption that each mix of the polysulphide rubber base and its catalyst was essentially identical, so that volume for volume, weights would be equal. This is one more limitation to be added to the set discussed previously (Turnbull, 1970, pp. 243–246) which must be kept in mind when making comparisons. These defects are not considered to be serious enough to render the estimates useless. Instead I believe that the estimates provide an adequate basis of comparison.

Table 1. *Thylacosmilus atrox* **Riggs weights (grams) and percentages of restored jaw muscles.**

Muscles	Weight		Jaw-closing musculature	Total jaw musculature
M. masseter complex		120.0	33.1	31.2
M. masseter	(94.5)		(26.1)	(24.6)
pars superficialis	45.3		12.5	11.8
pars profunda	49.2		13.6	12.8
M. zygomaticomandibularis	25.5		7.0	6.6
M. temporalis complex		208.1	57.5	54.1
M. temporalis	—		—	—
pars superficialis	121.6		33.6	31.6
pars profunda	86.5		23.9	22.5
M. pterygoideus complex		34.3	9.5	8.9
M. pt. internus	28.9		8.0	7.5
M. pt. externus	5.4		1.5	1.4
TOTAL JAW-CLOSING MUSCULATURE		362.4	100.1	94.2
M. digastricus complex		22.0		5.7
M. dig. posterior	16.3			4.2
M. dig. anterior	5.7			1.5
TOTAL JAW MUSCULATURE		384.4		99.9

Table 2. Percentages of jaw-closing muscle groups in *Thylacosmilus*, compared with those in some living generalized and specialized carnivorous forms (from Turnbull, 1970).

Genus	Muscle Complex		
	Masseter	Temporalis	Pterygoideus
Thylacosmilus	33.1	57.5	9.5
Didelphis	34.2	57.0	8.9
Echinosorex	26.9	61.2	11.8
Felis	35.2	54.3	10.5

Fig. 5 Outline drawings of the skull and jaws of *Thylacosmilus* in lateral view with the fields of attachment of each of the jaw-closing muscle groups indicated in red.

  A. The masseteric group.

  B. The temporal group.

  C. The pterygoid group.

  This outline follows the scheme used previously (Turnbull, 1970, fig. 42), where antagonistic lever-arms and their forces are shown. Resolution of the forces applied at the joint is given (vector diagrams).

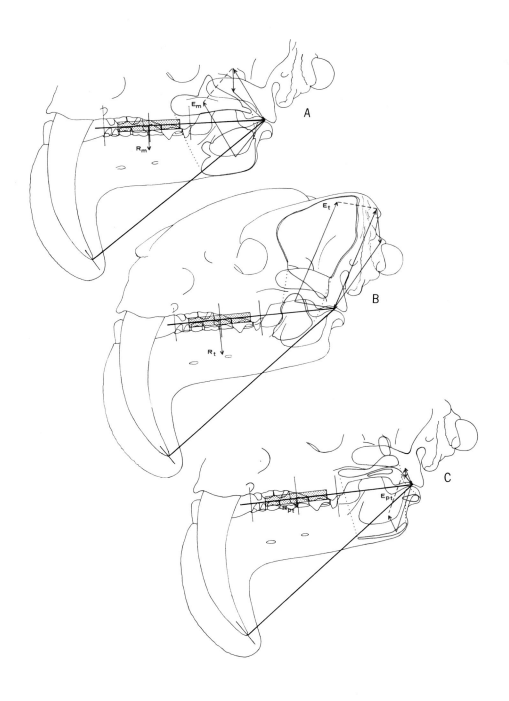

Proportions of the jaw-closing groups of muscles for *Thylacosmilus, Didelphis, Echinosorex* and *Felis* are compared in Table 2. The proportions of *Thylacosmilus* and *Didelphis* are almost identical, and in *Thylacosmilus* and *Felis* they are also very close. Comparison of *Thylacosmilus* with *Echinosorex* shows greater distinctness, but not so much as to cause difficulty in interpretation. My original presumption, that the muscle proportions of *Thylacosmilus* should fit the patterns of Specialized Group I (Carnivore-Shear Group) and of the Generalized Group (Turnbull, 1970, pp. 252–253 and 249 respectively), is confirmed.

Following the scheme set forth in my 1970 work (pp. 278–296) Figs. 5A, B, and C provide the working basis for calculations of the useful power (efficiency) as follows: on the outline of skull and jaws each muscle complex is superimposed in red and its lever system with effort and resistance lever arms and forces (red arrows) is shown. The useful power formula is applied: $E = M \times F_L \times F_X \times r$. Thus,

$$E_t = 57.42 \times 1 \times 1 \times .35 = 20.1 \text{ or } 55.5\%$$
$$E_m = 33.11 \times 1 \times 1 \times .40 = 14.2 \text{ or } 39.2\%$$
$$E_{pt} = 9.46 \times .50 \times 1 \times .41 = 1.9 \text{ or } 5.2\%.$$

Perhaps it should be noted that rarely does the direction of an angular pull depart far from that normal to its attachment on the jaw. In the cross-sectional plane, only the external pterygoid shows a direction that deviates by as much as 30° from a 90° angle of pull, and since it amounts to but one-sixth of the total of the pterygoid group, all of the cross-sectional correctional factors (F_X) approach values of 1. Hence, in addition to muscle mass, only the F_L factors and lengths of lever arms need be considered in *Thylacosmilus*.

Conclusion

Thylacosmilus, a most remarkable marsupial with highly specialized carnivorous dental adaptations, fits the pattern of living carnivorous mammals. It utilized a generalized masticatory musculature to operate its specialized dentition. The most striking dental feature, the hyperdeveloped canine, was powered by whole-body or other body movements, not primarily by the masticatory musculature. By these muscle restorations some flesh is put onto this interesting fossil, and the results appear functionally reasonable, so that the musculature as restored approximates that which the animal possessed in life.

Acknowledgments

I benefited from the help of a number of people. Tibor Perenyi advised and helped with modelling the muscles, and drew the overlays for Figs. 1, 2, and 5. The drawings in Figs. 1 and 2 are from Rigg's (1934) monograph. John Harris cast each modelled Plastilina muscle in polysulphide rubber. John Baylis photographed the cast of skull and jaws with the restored muscles (Figs. 3 and 4). Their assistance is gratefully acknowledged, as is that of Mary Alexander who typed the manuscript. My wife, Priscilla, and Rainer Zangerl have read the manuscript, discussed it and offered helpful comments and criticisms. Their help is most appreciated.

Literature Cited

ADAMS, L.A.
  1919   A memoir on the phylogeny of the jaw muscles in recent and fossil vertebrates. Ann. New York Acad. Sci., 28: 41–166, 13 pls.

BECHT, G.
  1953   Comparative biologic-anatomical researches on mastication in some mammals. I and II. Proc. Kon. Ned. Ak. v. Wet., ser. c, 56(4): 508–526, 14 figs., 1 pl., 3 tables.

BOHLIN, B.
  1940   Food habit of the machaerodonts, with special regard to *Smilodon*. Bull. Geol. Inst. Upsala. 27: 156–174, figs. 1–4.

MATTHEW, W.D.
  1910   The phylogeny of the Felidae. Bull. Amer. Mus. Nat. Hist. 28 (art. 26): 289–310, figs. 1–15.

MAYNARD SMITH, J. AND J.R.G. SAVAGE
  1959   The mechanics of mammalian jaws. School Sci. Rev., 141: 289–301, figs. 1–8.

RIGGS, E.S.
  1934   A new Marsupial sabertooth from the Pliocene of Argentina and its relationship to other South American Predaceous Marsupials. Trans. Amer. Phil. Soc., n.s., 14: 1–32, figs. 1–4, pls. 1–8.

SCHUMACHER, G.H.
  1961   Funktionelle Morphologie der Kaumuskulatur. Gustav Fischer Verlag, Jena. I–XVI, 1–262 pp., 138 figs., 16 diagrams.

SIMPSON, G.G.
  1941   The function of saber-like canines in carnivorous mammals. Amer. Mus. Novitates, 1130: 1–12, figs. 1–4.

TURNBULL, W.D.
  1970   Mammalian masticatory apparatus. Fieldiana: Geol., 18(2): 147–356, 48 figs., 5 tables.

Mammals of the Hand Hills Formation, Southern Alberta[1]

John E. Storer
Provincial Museum of Alberta

Abstract

Fossil mammals of two different ages occur in the Hand Hills Formation of southern Alberta: *Merychippus* sp., *Hipparion* sp., Camelidae, gen. et sp. indet., Sciurinae, gen. et sp. indet., *Copemys* sp., and *Pseudadjidaumo* cf. *russelli* are of latest Miocene or earliest Pliocene age; and Elephantidae, gen. et sp. indet., *Equus* cf. *conversidens*, *Spermophilus* cf. *richardsonii*, *Microtus* cf. *pennsylvanicus*, *Geomys* sp., and Leporidae, gen. et sp. indet., are of Irvingtonian or later age. The Hand Hills Formation comprises a single unit, including in places a layer of till, and contains a mixed mammalian fauna whose different temporal components are not separated by a stratigraphic marker.

Introduction

The age of the Hand Hills Formation, which lies at the summit of the Hand Hills, in Townships 29 and 30, Ranges 16 and 17 w of the fourth Meridian, has long posed a problem to geologists. L.S. Russell (1957, pp. 17–18) summarized early conclusions on the matter: "Tyrrell (1887), who was the first geologist to describe the area, correlated the conglomerate with that of the Cypress Hills, a natural conclusion from the data available. However, Warren (1939) pointed out that the elevation of the Hand Hills conglomerate was much too low for it to have been part of the sheet of gravel that must have extended eastward from the newly uplifted Rockies in Early Oligocene time, and of which the Cypress Hills conglomerate is a remnant. Warren suggested instead that the Hand Hills gravel and plain were equivalent to the Flaxville gravel and plain of northeastern Montana, the age of which is probably early Pliocene."

---

[1]Natural History Contribution No. 15, Provincial Museum of Alberta.

In 1956, Russell discovered an astragalus that he (1958) determined as an advanced type of horse, possibly the late Pliocene – early Pleistocene *Equus* (*Plesippus*). His find demonstrated that the Hand Hills Formation must be in part Blancan or younger.

Until recently, Russell's equid astragalus was the only identifiable specimen that had been recovered from the Hand Hills Formation. Since 1970, however, additional specimens have been recovered from a gravel pit in Legal Subdivision 16, Section 16, Township 30, Range 17 w fourth Meridian, on the northern-most extension of the Hand Hills. This pit is operated by the Municipal District of Starland No. 47, on land owned by Messrs John W. and Robert S. Sinclair. Although intensive collecting continues, a description of some specimens recovered so far will provide new evidence on the age of the Hand Hills Formation. A pre-liminary faunal list has already been published (Storer, 1972).

Most of the specimens discussed here have been collected in bulldozed material, and are not stratigraphically controlled. Two suites of specimens come from a bulldozed cut (Fig. 1) near the southernmost edge of the pit. One is a collection made in 1970 and 1971 by Mr. D.B. Schowalter, then of the University of Alberta, from a site near the top of the exposed section (Univ. Alberta [UA] local-ity); some specimens were found in place in the till, and some were in gravel immediately above the till. In 1971 a screening collection was made by crews from the Provincial Museum of Alberta from gravel taken directly from the pit wall beneath the till (PMA locality). Specimens from the UA locality represent both the Mio-Pliocene and Pleistocene faunas; those from the PMA locality are all referable to the Mio-Pliocene fauna.

Gravel above the till is probably not in its original position: it shows no cross-bedding, or other structure typical of gravel in place, and has probably been moved into its present position during the working of the pit. At every other locality in the Hand Hills where I have seen the till exposed, it is overlain by soil. Thus it is possible that some specimens collected at the UA locality originated below the till.

Although the evidence summarized above tempts one to conclude that the lower gravel is Mio-Pliocene and the till Pleistocene, evidence from another gravel pit calls this conclusion into question. Russell collected his equid astragalus, assigned in this paper to the Pleistocene fauna, from a pit in which the till is not exposed; the specimen was distinctively coloured, and he could tell from what area of the pit wall it had probably come. The pit is about 7½ mi (12 km) from the one discussed in this paper, and lies at about an equal altitude; equivalent gravels should be present in both pits.

The glacier that deposited the till probably worked some Pleistocene material into the earlier gravel, and picked up Mio-Pliocene specimens, incorporating them into the till. Mio-Pliocene specimens from the UA locality are broken or abraded, and could have been reworked.

The gravel, which is of fluviatile origin and was derived from the west, was probably uplifted and eroded before glaciation. A disturbed zone several centi-metres thick demonstrates that the gravel-till contact is an unconformity. Distur-bance of the lower gravel that would explain the deposition of Pleistocene speci-mens has not been observed: it appears that Russell's horse astragalus was not reworked, but was in its original place of deposition in the gravel.

Fig. 1 Section, gravel pit in Lsd. 16, Tp. 30, Rge. 17 W. 4th Merid.: facing south, near
southernmost edge of pit (ruler = 46 cm).
  A. Gravel, coarse, mainly brown and gray quartzites and cherts, brown sand matrix,
     no sorting or crossbedding evident, 23–46 cm.
  B. Till, gray, white specks, inclusions cobbles, 25–58.5 cm.
  C. Gravel, coarse, mainly brown and gray quartzites and cherts, with marl balls, sand
     lenses, brown to gray sandy matrix, cemented in place, weakly sorted, weakly cross-
     bedded, about 200 cm to base of section; (a) level of UA locality; (b) level of PMA
     locality.

I must conclude that the gravel contains a mixed mammalian fauna whose temporal components are not separated by a stratigraphic marker. The till also contains a mixed fauna, and is probably Pleistocene. I consider the Hand Hills Formation a single unit, and include the till within it.

Procedures

Specimens were measured with vernier calipers reading to 0.1 mm, and all measurements are given in millimetres. The following abbreviations are used: NMC: National Museums of Canada, National Museum of Natural Sciences, Vertebrate Paleontology Collections; PMA P: Provincial Museum of Alberta, Paleontology Collections; PMA Z: Provincial Museum of Alberta, Zoology Collections; UA: University of Alberta, Department of Geology, Vertebrate Paleontology Collections.

Systematics

Class Mammalia
Order Proboscidea
Family Elephantidae

Elephantidae, gen. et sp. indet.

Material
Enamel fragments, UA 8515, UA 8516.

Provenance
UA locality.

Description
Two large dental fragments, with enamel 2.1 to 2.8 mm thick, represent a proboscidean. In one (UA 8515), the enamel bends into a U-shape enclosing cementum, suggesting the area between grinding ridges of an elephantid cheek tooth.

Discussion
The fragments seem definitely referable to an elephantid, because they represent a hypsodont, strongly lophate, rather than a brachydont, cuspidate tooth. Elephantids make their first appearance in North America in the Irvingtonian (Hibbard *et al.*, 1965, table 2).

Order Perissodactyla
Family Equidae

Merychippus sp.
Fig. 2A, B

Material
L?P^4, UA 8517. R?M$_1$, UA 8518. RM$_3$, UA 8519. Metacarpal III, UA 8520. Median or proximal phalanx III, UA 8521.

Provenance

UA locality, bulldozed areas.

Description

A portion of the distobuccal region of a left upper cheek-tooth (Fig. 2A), possibly P⁴, shows the hypofossette extending to a point about 10 mm above the base of the crown. The crown's dorsoventral diameter is 30.3 mm. No pli hypostyle is present, and the pli postfossette is three-pointed.

A right lower cheek-tooth (Fig. 2B), possibly M_1, is relatively complete. The protostylid has been removed by interdental wear, but it appears to have been thin buccolingually at the top of the crown, expanding basally. Pre- and postfossettids are simple and open nearly to the base of the crown. The mesostylid and metaconid are well separated by a deep, cement-filled entostriid that is open to the base of the crown. Length from the mesial end of the metaconid to the distal end of the mesostylid is 9.5 mm, and the dorsoventral height of the crown is 30.8 mm; accurate mesiodistal and buccolingual measurements cannot be taken because the hypostylid and most of the cementum are broken away.

A small fragment of M_3, UA 8519, shows a simple postfossettid open far down the crown. The column is broken, and the base of the crown is not preserved.

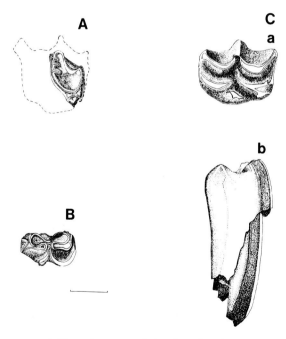

Fig. 2 *Merychippus* sp. and *Hipparion* sp. teeth (scale = 1 cm).
    A. *Merychippus* sp. L?P⁴ UA 8517, occlusal aspect, mesial to left, buccal to top of page.
    B. *Merychippus* sp. R?M_1 UA 8518, occlusal aspect, mesial to left, buccal to top of page.
    C. *Hipparion* sp. LM¹ PMA P72.15.1; (a) occlusal aspect, mesial to left, buccal to top of page; (b) distal aspect, buccal to right, occlusal to top of page.

Table 1. Measurements (mm) of postcranial elements of Hand Hills *Merychippus* compared with corresponding measurements of elements from the upper Hemingfordian–lower Barstovian Mascall fauna (from Downs, 1956, table 20).

Element	Hand Hills	Mascall
Metacarpal III, distal end		
anteroposterior diameter	17.3	17.3–17.5
transverse diameter	21.4	21.2–22.3
Median or proximal phalanx III	23.6	19.3–22.6 (median)
transverse diameter, proximal end	—	18.6–19.7 (proximal)

A fragment of the distal end of metacarpal III, UA 8520, is water-worn but still identifiable. It shows the anteroposterior flattening of the shaft characteristic of the median metacarpal.

UA 8521 may represent the proximal end of a median or proximal phalanx III. Only a transverse measurement of the proximal end can be taken. It cannot be ascertained whether the element is from the manus or pes.

Discussion

Although no mesiodistal or buccolingual measurements of crowns of the Hand Hills teeth are available, it is clear that all three specimens compare well in size and pattern to teeth of *Merychippus*, e.g. *M. isonesus* (Osborn, 1918, pls. 13, 14), a characteristic Hemingfordian-Barstovian equid. Crown heights, and the depth of the hypofossette in P^4 are comparable to those of *Merychippus* cf. *M. isonesus* from the Wood Mountain Formation (Storer, 1970b, pp. 117, 119, 121).

The distal fragment of metacarpal III compares well in size with those studied by Downs (1956), as does the fragment of phalanx (Table 1).

Merychippus ranges from middle Miocene to lowermost Pliocene. The five specimens considered here are of consistent size, and all could represent the same species. No specific determination can be made, because the material is fragmentary.

Hipparion sp.
Fig. 2c

Material
LM¹, PMA P72.15.1.

Provenance
Bulldozed area.

Description
The tooth (Fig. 2c, a and b) is nearly unworn. It measures 21.8 mm mesiodistally, 19.0 mm buccolingually, and 43.2 mm dorsoventrally. The hypofossette is open

broadly for about 25 mm ventrally from its origin, down the entire preserved portion of the distal side: it presumably was open nearly to the base of the crown, another 12 mm, and shows no sign of narrowing as does the hypofossette of P⁴, UA 8517 of *Merychippus* sp. Details of the pre- and postfossettes are not developed. All sides of the crown show cement.

Discussion

The tooth is referred to *Hipparion* because of its great hypsodonty, and because the hypofossette appears to have extended nearly to the base of the crown. Comparison can be made to *Hipparion* sp. from the Wood Mountain Fauna (Storer, 1970b, p. 127, table 21), which it resembles in size, height of crown, and depth of hypofossette.

The temporal range of *Hipparion* is latest Miocene (Storer, 1970b) to early Pleistocene.

Equus* cf. *conversidens
Fig. 3

Material

RI¹, UA 8523. RP², UA 8524. LM₁, PMA P71.54.1. LM₂, PMA P72.19.1. Left lower cheek tooth, UA 8525, UA 8526, PMA P71.37.1. Right lower cheek-tooth, UA 8527.

Provenance

UA locality, bulldozed areas.

Description

The incisor (Fig. 3A, a and b) is assigned to the upper dentition because the curvature of its crown compares with those from Cochrane, Alberta, discussed by Churcher (1968, pp. 1474–1475). Its crown is slightly wider buccolingually than long mesiodistally, the root is closed, and the fossette is large and filled with cementum.

A specimen representing P² (Fig. 3B), UA 8524, is badly broken, and dependable measurements cannot be taken: it is estimated that if the tooth were unbroken it would measure 38 mm mesiodistally and 30 mm buccolingually. Much of the protocone is broken away: it extended well distally from its point of strong attachment to the protoloph. The prefossette loop does not extend lingually as far as the borders of the pre- and postfossettes. There is a double pli protoloph, and the pli caballine is medium in size.

The pli prefossette is shallow and single, with a weak plication buccal to it; the pli postfossette is deep and single. The metaloph is thin near the metacone. The pli hypostyle is small, the hypofossette is filled with cementum and appears closed, and thick cement is present on all but the distal side of the tooth.

A fragment of LM₁ (Fig. 3C, a and b), PMA P71.54.1 is too badly broken for measurements to be taken. The entostylid is well rounded, and the mesostylid has a deep distal groove. There appears to have been a deep entostriid, and the postfossettid is long and broad.

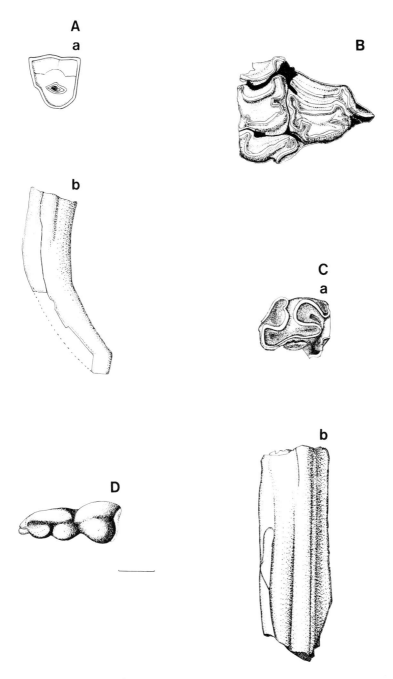

Fig. 3 *Equus* cf. *conversidens* teeth (scale = 1 cm).

    A. RI^1 UA 8523; (a) occlusal aspect, mesial to right, buccal to top of page; (b) distal aspect, buccal to left, occlusal to top of page.

    B. RP^2 UA 8524, occlusal aspect, mesial to right, buccal to top of page.

    C. LM_1 PMA P71.54.1; (a) occlusal aspect, mesial to right, buccal to top of page; (b) buccal aspect, mesial to left, occlusal to top of page.

    D. LM_2 PMA P72.19.1, occlusal aspect, mesial to right, buccal to top of page.

The well-preserved LM_2 (Fig. 3D), PMA P72.19.1, is broken mid-height. It measures 30.2 mm mesiodistally, and 12.4 mm buccolingually. The tooth is unworn, and traces of thin cement remain on all enamel surfaces. Both the mesostylid and metaconid are oval in section, and the entostriid is broad, deepening toward the base of the crown. The prefossettid is long mesiodistally, and broadest transversely at its mesial end; the postfossettid is long mesiodistally.

Fragments of lower cheek teeth represent the areas of the postfossettids, and specimens PMA P71.54.1, PMA P71.37.1, and UA 8526 have postfossettids 15.0, 11.7, and 10.5 mm in mesiodistal diameter, respectively.

Discussion

The upper incisor, UA 8523, is similar in size (Table 2) and pattern to those of *Equus conversidens* illustrated by Churcher (1968, fig. 9). It fits well within the range of variation of *E. conversidens*.

It is estimated that the Hand Hills P^2 is 20-30 per cent larger (Table 3) than corresponding teeth of *E. conversidens* discussed by Dalquest and Hughes (1965, table 1). This difference in size is, however, not sufficient to exclude the tooth from *E. conversidens*. Similarities in crown pattern to teeth of *E. conversidens* include the medium-sized pli caballine, the thin metaloph near the metacone, and the relatively simple fossettes, with small pli hypostyle and simple pli pre- and postfossettes.

A fragment of M_1, PMA P71.54.1, compares well in size to the molars from Slaton Quarry, Texas and Tajo de Tequixquiac, Mexico, illustrated by Churcher (1968, fig. 8 c,d). The distal groove of the mesostylid and the deep entostriid are paralleled in the Mexican specimen; the rounded entostylid and relatively long and broad postfossettid differ from both of Churcher's specimens. These latter differences may represent individual variation: on the basis of similar size and occlusal pattern, the fragment of lower molar from the Hand Hills is close to *Equus conversidens*. Other fragments of lower cheek teeth compare well in size to PMA P71.54.1, and appear to belong to the same species.

The M_2, PMA P72.19.1, is longer mesiodistally and narrower buccolingually than M_2's reported by Dalquest and Hughes (1965, table 2). The tooth is unworn, however, and interdental wear would remove the hypostylid, considerably shortening the mesiodistal dimension. Maximum buccolingual width would occur near the base of the crown, which is not preserved on the Hand Hills specimen. PMA P72.19.1 cannot be excluded from *E. conversidens* on the basis of size.

The Hand Hills astragalus NMC 9449, described by Russell (1958, pp. 1–3) was not inspected. But the specimen agrees in size (Table 4) and characteristics with NMC 12444, the astragalus of *Equus conversidens* from Pashley, Alberta (Churcher and Stalker, 1970, table 1, fig. 3), and cannot be excluded from *E. conversidens*.

Equus specimens described here, and Russell's Hand Hills equid astragalus, compare closely with *E. conversidens*. The known record of *E. conversidens* ranges from Yarmouthian to Holocene (Churcher, 1968, p. 1486; Churcher and Stalker, 1970, pp. 1024–1025).

Table 2. Measurements (mm) of the upper incisor from the Hand Hills (UA 8523) compared with those of *Equus conversidens* from Cochrane, Alberta (from Churcher, 1968, table IV).

Measurement	Hand Hills	Cochrane
Mesiodistal diameter of occlusal surface	14.5	13.9–22.2
Buccolingual diameter of occlusal surface	13.2	11.4–12.5
Direct length from occlusal buccal enamel margin to tip of root	49.1	56.0–68.7
Mesiodistal diameter of enamel fossette	8.3	2.2–12.5
Buccolingual diameter of enamel fossette	5.2	1.8–6.0

Table 3. Estimated measurements (mm) of P^2 from Hand Hills (UA 8524) compared with those of *Equus conversidens* from Texas and Mexico (from Dalquest and Hughes, 1965, table I).

Measurement	Hand Hills	Texas, Mexico
Mesiodistal diameter	38	31.7–33.3
Buccolingual diameter	30	20.7–23.6

Table 4. Measurements (mm) of Hand Hills equid astragalus, NMC 9449 (from Russell, 1958, p. 3 and estimated from figs. 1, 2) compared with those of *Equus conversidens* astragalus from Pashley, Alberta, NMC 12444 (from Churcher and Stalker, 1970, table I).

Maximum Measurement	Hand Hills	Pashley
Anteroposterior diameter	56.7	52.3
Transverse diameter of tibial trochlear	51 (est.)	50.1
Transverse diameter of navicular facet	53 (est.)	51.5
Proximodistal length between facets	61	57.5

Order Artiodactyla
Family Camelidae

Camelidae, gen. et sp. indet.
Fig. 4

Material
LM$_1$ or M$_2$, PMA P71.42.1. Proximal phalanx, PMA P70.22.1.

Provenance
Top of section, north end of pit; bulldozed area.

Description
The fragment of a lower molar (Fig. 4A) is broken longitudinally, and is from a tooth about 24.9 mm in mesiodistal length, with a height of crown of 30.8 mm. The hypoconid is more angular buccally than is the protoconid. No basal stylid blocks the ectoflexid, and no cingulum is visible.

The phalanx (Fig. 4B) is proportionately long and slender, and its measurements are given in Table 5.

Discussion
The phalanx compares in slenderness to those of modern *Camelus* as do early Pliocene examples studied by Webb (1969, p. 159): "There is a remarkably close resemblance in skeletal features between *P.* [*Procamelus*] *grandis* and *Camelus bactrianus*. The main difference is that most elements of *P. grandis* are much lighter in build even though they agree with the corresponding bones of the modern species in length." In Table 5 it can be seen that the measurements of the Hand Hills camelid phalanx agree proportionally far more closely to those of phalanges of *Procamelus grandis*, a Clarendonian camelid, than to corresponding measurements of the large Blancan *Gigantocamelus spatula*, the Pleistocene

Table 5. Measurements (mm) of proximal phalanx, Hand Hills camelid (PMA P70.22.1) compared with those of proximal phalanges of *Procamelus grandis* (from Gregory, 1942, p. 376), *Gigantocamelus spatula* (from Meade, 1945, p. 534), *Camelops huerfanensis* (from Hay, 1927, p. 100), and *Camelus dromedarius* (from Hay, 1927, p. 103).

Measurement	Hand Hills Camelid	*Procamelus grandis*	*Gigantocamelus spatula* (medians*)	*Camelops huerfanensis*	*Camelus dromedarius*
Proximodistal length (A)	76.0	88.8	112–138	119	94
Transverse diameter of proximal end (B)	23.9	28.9	55–62.5	45	41
Transverse diameter of distal end (C)	17.5	22.5	45–52	36	34
A ÷ B	3.2	3.1	2.1*	2.6	2.3
A ÷ C	4.3	3.9	2.6*	3.3	2.8

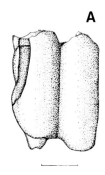

Fig. 4 Camelidae, gen. et sp. indet. tooth and phalanx (scales = 1 cm).
  A. LM_1 or M_2 PMA P71.42.1, buccal aspect, mesial to left, occlusal to top of page.
  B. Proximal phalanx PMA P70.22.1, anterior aspect, distal to top of page.

Camelops huerfanensis, or the modern *Camelus dromedarius*. It is more likely that the phalanx represents a Miocene or early Pliocene camelid than a Pleistocene one.

The portion of lower molar is of a size consistent with its assignment to the same species as the phalanx, and probably is referable to the Miocene-Pliocene camelid also.

Order Rodentia
Family Sciuridae
Subfamily Sciurinae

Spermophilus cf. *richardsonii*
Fig. 5A–E

Material
RP^4, UA 8528, UA 8529. LM^1 or M^2, UA 8530. RP_4, UA 8531. LM_2, UA 8532. RM_3 UA 8533. R. humerus, PMA P72.15.9.

Provenance
UA locality, except PMA P72.15.9 from bulldozed area.

Description
Upper cheek teeth (Figs. 5A,B) are proportionally wide buccolingually and short mesiodistally, and have a generally triangular shape. The teeth are high-crowned, and in general the protolophs and metalophs are high.

In an upper cheek tooth, the protocone is crescentic, and is connected to the paracone by a high, straight protoloph running buccolingually; the protoconule is not visible, and may not have been present. The metaloph is strong and slightly constricted at its junction with the protocone; the metaconule is prominent, and is nearly as large as the metacone. A strong but low posterior cingulum has a slightly thickened origin just distal to the point of junction of the protocone and metaloph, and runs to a point distolingual to the base of the metacone.

In the P^4's at hand (Fig. 5A), the parastyle is near one end of an arcuate anterior cingulum that encloses a basin which is longer mesiodistally than those of the molars. No mesostyle is present.

The M^1 or M^2 (Fig. 5B) shows a reduced mesiobuccal corner. The parastyle is large, and from it runs a strong anterior cingulum, directed nearly lingually and bearing a large cuspule mesial to the base of the protocone. The mesostyle is low and oval.

The P_4 (Fig. 5C) is as broad buccolingually as it is long mesiodistally, and the M_1 or M_2 (Fig. 5D) is much broader than its length. The P_4 shown in Fig. 5C has a deep trigonid basin, opening to the mesial surface of the crown just lingual to the large anteroconid; the M_1 or M_2 (Fig. 5D) is far more worn, but it appears that the anteroconid was small or absent. Mesostylids in P_4 and M_1 are low, and entoconids are not clearly separated from posterolophids. Small mesoconids are present on the ectolophids.

In M_3 (Fig. 5E), the metaconid is high and prominent, and the protoconid is worn into a crescent. The prominent metalophid ends sharply about midway between the metaconid and protoconid, so that the trigonid basin connects lingually with the talonid basin. An ectolophid bears a small mesoconid near its mesial end, just distal to the protoconid. The entoconid and mesostylid are not clearly separated from the posterolophid; the mesostylid is separated from a cingulum, running distally from the metaconid, by a deep valley which runs mesiobuccally. The talonid basin is marked by a "basin trench" (Repenning, 1962, p. 544) at its buccal, mesial, and distal sides. Enamel in the lingual portion of the talonid basin is rugose.

The distal end of a right humerus, PMA P72.13.9, is damaged on its outer edge. It compares well to humeri of *Spermophilus richardsonii*, but is slightly broader transversely than examples from the PMA Zoology Collections.

Discussion

The teeth and humerus represent an advanced spermophile showing great similarity to *Spermophilus richardsonii*, a species known from Irvingtonian to present (Hibbard, 1970, table 4). Although more nearly complete material will be necessary for precise identification, similarities in sizes of the teeth (Table 6), the basin trench in M_3, positions and sizes of mesoconids in lower cheek teeth, and general features of upper cheek teeth are extremely close.

Fig. 5 Sciuridae, teeth in occlusal aspect (scale = 1 mm). Mesial to left in B, C, E, F; to right in A, D; buccal to top of page.
    A-E. *Spermophilus* cf. *richardsonii*.
    A. RP^4 UA 8528.
    B. LM^1 or M^2 UA 8530.
    C. RP_4 UA 8531.
    D. LM_2 UA 8532.
    E. RM_3 UA 8533.
    F. Sciurinae, gen. et sp. indet. LdP^4 PMA P71.54.36.

A **B**

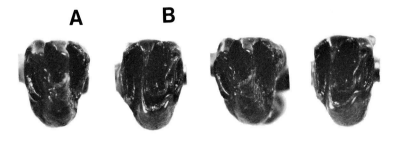

D

C

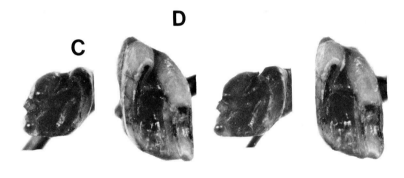

E

F

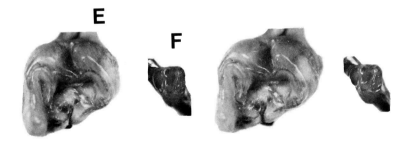

Table 6. Measurements of diameters (mm) of teeth of *Spermophilus* cf. *S. richardsonii* from the Hand Hills compared with those of a modern specimen of *S. richardsonii*, PMA Z70.36.3.

Tooth	Hand Hills		Tooth	PMA Z70.36.3	
	Mesiodistal	Buccolingual		Mesiodistal	Buccolingual
P⁴ UA 8528	2.3	3.0	P⁴	2.1	2.8
UA 8529	1.9	2.5			
M¹ or M² UA 8530	2.1	3.0	M¹	2.0	3.0
			M²	2.0	3.0
P₄ UA 8531	2.3	2.3	P₄	2.1	2.5
M₁ or M₂ UA 8532	2.7	3.2	M₁	2.3	2.8
			M₂	2.4	2.9
M₃ UA 8533	3.5	3.1	M₃	3.6	2.9

Sciurinae, gen. et sp. indet.
Fig. 5F

Material
LdP⁴, PMA P71.54.36.

Provenance
PMA locality.

Description

The tooth (Fig. 5F) has a roughly triangular crown and is slightly broader buccolingually (1.4 mm) than long mesiodistally (1.2 mm). Its crown is short dorsoventrally, and has no roots; only an enamel cap is preserved.

An anterior cingulum forms a loop from the mesial edge of the protocone to the mesiobuccal edge of the paracone, enclosing a large mesial shelf that slopes distally. Both protoloph and metaloph are complete; the metaloph bears a small metaconule distobuccal to the distal end of the protocone. The paracone is slightly larger than the metacone, and a small mesostyle lies between them. The posterior cingulum is large but low, running from the distal edge of the protocone to the distal edge of the metacone and enclosing a narrow valley.

Discussion

No generic assignment can be given to this small sciurine. The specimen is from the lower-level screening sample, and is of the same colour and preservation as teeth of *Copemys* sp. and *Pseudadjidaumo* cf. *P. russelli*: it presumably represents one of the Miocene-Pliocene taxa in the deposit.

Family Cricetidae
Subfamily Cricetinae

Copemys sp.
Fig. 6A, B

Synonymy
Peromyscus sp. (Storer, 1972, p. 120)

Material
RM¹, PMA P71.54.37. RM₁, PMA P71.54.38.

Provenance
PMA locality.

Description
The RM¹ (Fig. 6A) is badly worn, and the mesial end is missing. The posterior cingulum is well separated from the distal end of the metacone (terminology from Lindsay, 1972, fig. 40), and the mesostyle and mesoloph are strong. No enterostyle is present in the lingual cingular shelf. The hypocone is set farther mesially than the metacone, and thus good alternation of cusps is present. However, protolophule II is directed lingually, and not towards the hypocone. Maximum buccolingual diameter is 1.2 mm.

The RM₁ (Fig. 6B) bears two cusps on the anterior cingulum, and the anterolophulid runs mesiobuccally from the junction of the protoconid and metaconid, to the distolingual edge of the buccal cusp of the anterior cingulum. The posterior cingulum curves mesiolingually to the distolingual edge of the entoconid, and the hypoconid projects buccally at the distal end of the tooth, well past the rest of the buccal margin. Alternation of cusps is well developed; however, the entolophid runs directly buccally from the entoconid, meets the posterior mure just mesial to the hypoconid, and does not line up with the anterior mure. Lingual and buccal valleys distal to the anterior cingulum are blocked by low cingula, and no other accessory cusps or lophs are present on the tooth. The tooth is proportionally long mesiodistally and narrow buccolingually and is broken into mesial and distal portions: in Fig. 6B, the pieces are placed together, but the break is clearly visible. The estimated mesiodistal diameter is 1.5 mm, and the buccolingual diameter at the hypoconid and entoconid is 0.9 mm.

Discussion
These two teeth are referable to *Copemys* by Lindsay's (1972, p. 75) definition, because in M¹ protolophule II does not line up with the mesial arm of the hypocone, and because M₁ is proportionally long and slender, and its entolophid does not line up with the distal arm of the protoconid (anterior mure). The known temporal range of *Copemys* (Lindsay, 1972, p. 74) is Barstovian to Hemphillian (late Miocene to late Pliocene).

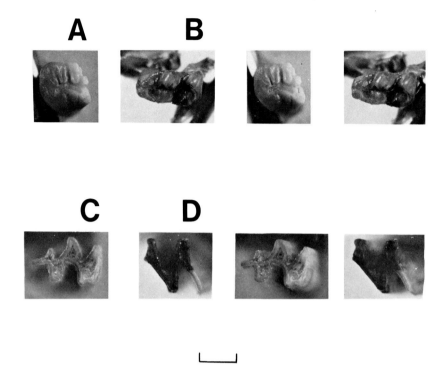

Fig. 6 Cricetidae, teeth in occlusal aspect (scale = 1 mm). Mesial to left in B, D; to right in
A, C; buccal to top of page.
A-B. *Copemys* sp.
A. RM¹ PMA P71.54.37.
B. RM₁ PMA P71.54.38.
C-D. *Microtus* cf. *M. pennsylvanicus*.
C. RM² UA 8534.
D. RM₁ or M₂ UA 8535.

Subfamily Microtinae

Microtus cf. *pennsylvanicus*
Fig. 6C, D

Material
RM², UA 8534. RM₁ or M₂, UA 8535.

Provenance
UA locality.

Description
The RM² (Fig. 6C) has a lingually rounded mesial loop, and a dentine tract causes
an interruption in the enamel of the lingual side of the mesial loop. Two triangles
distal to the mesial loop are closed and point mesiolaterally; reëntrant angles are
sharp. The last triangle and the distal loop are broken, but it is apparent that the
former was closed, and that the distal (fifth) loop was small. This tooth is unrooted,

and small amounts of cementum are present. The mesiodistal diameter is 2.0 mm, the buccolingual diameter 1.2 mm.

Three triangles from the distal end of a microtine lower molar (Fig. 6D) are also preserved. The distal loop has a buccal enamel interruption. All triangles are closed, and sharply pointed mesiolaterally. Reëntrant angles are sharp, and cementum is present. The tooth fragment is unrooted, and the buccolingual diameter is 1.4 mm.

Discussion

Specimens from the Hand Hills compare well with corresponding teeth of *Microtus pennsylvanicus* (Hibbard, 1956, text fig. 1; Schultz, 1965, fig. 2E). Schultz (1965, p. 253) mentions the presence of a rounded fifth loop in M^2 as a characteristic of *M. pennsylvanicus*; the Hand Hills specimen (Fig. 6C) probably possessed this feature. The temporal range of *Microtus* is Irvingtonian to present (Hibbard *et al.*, 1965, fig. 2).

Family Eomyidae

Pseudadjidaumo cf. *russelli*
Fig. 7A

Synonymy
Adjidaumo cf. *A. russelli* (Storer, 1972, p. 120).

Material
LM_1 or M_2, PMA P71.54.39.

Provenance
PMA locality.

Description

This specimen (Fig. 7A) is the lingual portion of a left lower molar. Only the distal root is complete, but a fragment of root beneath the metaconid indicates that the tooth was three-rooted.

The entoconid and metaconid are subequal in size, with the metaconid the higher of the two; the metaconid connects mesially to the anterior cingulum, and the posterior cingulum originates about midway between the entoconid and the posterior arm of the ectolophid. No mesolophid is present. The mesiodistal diameter of the crown is 0.9 mm.

Discussion

This tooth has, unfortunately, been lost; photographs and measurements were made, and the specimen was studied before its disappearance.

The fragment of tooth is about 0.1 mm longer mesiodistally than M_1 or M_2 of the holotype of "*Adjidaumo russelli*" (Storer, 1970a, pp. 1125, 1127, fig. 1), and differs from those teeth in showing a connection between the metaconid and anterior cingulum. In all other respects, the resemblance is close, especially in the absence of a mesolophid. Of the late Tertiary eomyids, this specimen is closest to "*A. russelli*," suggesting an uppermost Miocene or lowermost Pliocene age for this component of the Hand Hills Fauna.

It is certain that "*Adjidaumo russelli*" belongs in the genus *Pseudadjidaumo* Lindsay, which he diagnosed (1972, p. 36) as a "Small, brachydont rodent with transverse lophules and lophulids well developed; anterior and posterior valleys narrow and shallow, usually closed laterally except in M^3; anterior cingulum of lower molars usually uniting with anterior arm of protoconid near median plane of tooth." Observable characters of the Hand Hills tooth fit this diagnosis, and its proper designation is *Pseudadjidaumo* cf. *russelli*.

Family Geomyidae

Geomys sp.
Fig. 7B

Material
LI^1, UA 8536.

Provenance
UA locality.

Description
The tooth (Fig. 7B) represents the only geomyid that has been recovered from the Hand Hills Formation. Its mesiodistal diameter is 1.6 mm, and its buccolingual diameter 1.7 mm. Two longitudinal sulci are present: the more mesial is set close to the mesial edge of the tooth, and is shallow but clearly defined; the more distal is deep, and is just distal to the middle of the column.

Discussion
Two longitudinal sulci are characteristic of the upper incisor of *Geomys*, a genus whose first appearance is in the earliest Blancan (Hibbard *et al.*, 1965, table 2).

Order Lagomorpha
Family Leporidae

Leporidae, gen. et sp. indet.
Fig. 7C, D

Material
Upper molariform tooth, UA 8537, UA 8538, UA 8539. RP_4 or M_1 or M_2, UA 8540. RM_3, UA 8541.

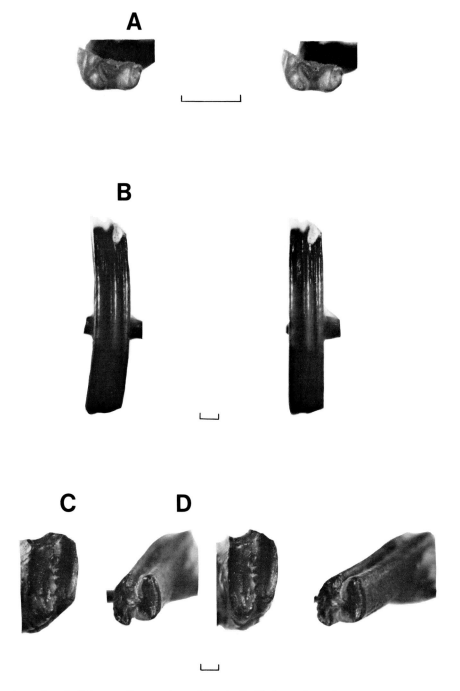

Fig. 7 *Pseudadjidaumo, Geomys*, Leporidae, teeth (Scales = 1 mm).
    A. *Pseudadjidaumo* cf. *russelli* LM_1 or M_2 PMA P71.54.39, occlusal aspect, mesial to right, buccal to top of page.
    B. *Geomys* sp. LI^1 UA 8536, buccal aspect, mesial to right, occlusal to top of page.
    C-D. Leporidae, gen. et sp. indet.
    C. P^4 or M^1 or M^2 UA 8537, occlusal aspect, buccal to top of page.
    D. RP_4 or M_1 or M_2 UA 8540, occlusal aspect, mesial to left, buccal to top of page.

Table 7. Measurements of diameters (mm) of teeth referred to Leporidae, gen. et sp. indet.

Upper molariform teeth	Mesiodistal	Buccolingual	
UA 8537	2.3	3.6	
UA 8538	2.7	4.4	
UA 8539	2.3	4.0	
	Mesiodistal	Buccolingual, trigonid	Buccolingual, talonid
P_4 or M_1 or M_2 UA 8540	2.0	1.9	1.8
M_3 UA 8541	1.4	1.4	1.3

Provenance

UA locality.

Description

In upper molariform teeth (Fig. 7c), the lingual hypostria is crenulated, and traverses half to two-thirds of the buccolingual width of the tooth. Cement is confined to the hypostria and the deep buccal fold.

In the P_4, M_1 or M_2 (Fig. 7D), the trigonid is shorter mesiodistally than the talonid and sharply pointed buccally. The talonid is rounded laterally and strongly connected to the trigonid lingual of centre.

In the M_3 (not shown), the trigonid is nearly as long mesiodistally as the talonid, and is slightly pointed buccally. The talonid is rounded lingually and connected to the trigonid just lingual of centre by a thick mesiodistal process. Cement is confined to the lingual and buccal folds. The trigonid and talonid are subequal in buccolingual width.

Discussion

Although it is certain that the specimens represent leporids, P^2 and P_3, the key teeth in lagomorph taxonomy, are not present in the samples recovered, and no definite assignment can be made. Table 7 lists the measurements of the various teeth.

Preservation is more like that of *Spermophilus, Microtus,* and *Geomys* teeth than that of the Miocene-Pliocene teeth. For this reason the Hand Hills leporid is grouped with the Pleistocene taxa.

Conclusions

The 12 mammalian taxa identified from the Hand Hills Formation can be divided into two groups. Known temporal ranges of Mio-Pliocene taxa (Table 8) overlap to give a late Barstovian age (latest Miocene), although *Pseudadjidaumo* may extend into the earliest Pliocene; ranges of the Pleistocene taxa (Table 9) indicate an Irvingtonian (late Kansan) or younger age.

It would appear from the ages of the two faunal components from the Hand Hills Formation that two widely separated episodes of deposition occurred, one in the latest Miocene or earliest Pliocene and the other in the Irvingtonian or later. Field work has revealed no stratigraphic marker between Mio-Pliocene and Pleistocene deposits in the Hand Hills.

Fossils representing the Mio-Pliocene portion of the deposit were not originally reported, and the Irvingtonian or later date attached to the Pleistocene portion of the Hand Hills Conglomerate is far later than the previously accepted Blancan age (Russell, 1958).

Table 8. Mio-Pliocene fauna of the Hand Hills Formation.

Taxon	Known temporal range
Merychippus sp.	Hemingfordian to Clarendonian
Hipparion sp.	Barstovian to Blancan
Camelidae, gen. et sp. indet.	—
Sciurinae, gen. et sp. indet.	—
Copemys sp.	Barstovian to Hemphillian
Pseudadjidaumo cf. *russelli*	Late Barstovian

Table 9. Pleistocene fauna of the Hand Hills Formation.

Taxon	Known temporal range
Elephantidae, gen. et sp. indet.	Irvingtonian to Holocene
Equus cf. *conversidens*	Irvingtonian to Holocene
Spermophilus cf. *richardsonii*	Irvingtonian to Present
Microtus cf. *pennsylvanicus*	Irvingtonian to Present
Geomys sp.	Blancan to Present
Leporidae, gen. et sp. indet.	—

Acknowledgments

Special thanks are due Mr. D.B. Schowalter, who collected much of the material described here, and was the source of valuable information about the deposit. Dr. R.C. Fox, University of Alberta, allowed me to describe material in the University's collection. Mr. Henry Gall, of Hanna, Alberta, generously donated *Equus* molar PMA P72.19.1.

Discussions with Mr. M.F. Skinner, American Museum of Natural History, Dr. G.E. Schultz, West Texas State University, and Drs. C.S. Churcher and L.S. Russell, University of Toronto and Royal Ontario Museum, provided ideas.

Mr. D.J. Merritt, Secretary of M.D. Starland, and Messrs John and Robert Sinclair were most co-operative in allowing PMA field crews to work in the gravel pit. Able field assistance was supplied by Messrs Stephen Holloway, Ronald Solkoski, Michael Wilson, Robert Bruinsma, and Alan Brunelle and Mrs. Barbara Storer.

Messrs D.A.E. Spalding, D.A. Taylor, and H.C. Smith and Miss J.O. Hrapko, of the Natural History Division, Provincial Museum of Alberta, read the manuscript and offered helpful comments. Fig. 1 was photographed by Mr. Stephen Holloway. The drawings presented here were made by Mr. Ronald Solkoski, and Miss Vivian Thierfelder made a series of drawings that were used extensively in the study of the specimens. Miss Elinor Huston typed the manuscript.

Mr. Ronald Solkoski and Mr. Kevin Williams are responsible for the time-consuming, tedious sorting of concentrated matrix. Their careful work contributed greatly to this project.

Literature Cited

CHURCHER, C.S.
  1968   Pleistocene ungulates from the Bow River gravels at Cochrane, Alberta. Can. J. Earth Sci., 5(6): 1467–1488.

CHURCHER, C.S. AND A. MACS. STALKER
  1970   A late, postglacial horse from Pashley, Alberta. Can. J. Earth Sci., 7(3): 1020–1026.

DALQUEST, W.W. AND J.T. HUGHES
  1965   The Pleistocene horse, *Equus conversidens.* Am. Midl. Nat., 74(2): 408–417.

DOWNS, T.
  1956   The Mascall fauna from the Miocene of Oregon. Univ. Calif. Publs. Geol. Sci., 31(5): 199–354.

GREGORY, J.T.
  1942   Pliocene vertebrates from Big Spring Canyon, South Dakota. Univ. Calif. Publs. Geol. Sci., 26(4): 307–446.

HAY, O.P.
  1927   The Pleistocene of western North America and its vertebrated animals. Publ. Carneg. Inst. Wash., 322B: 1–346.

HIBBARD, C.W.

1956   *Microtus pennsylvanicus* (Ord) from the Hay Springs local fauna of Nebraska. J. Paleont., 30(5): 1263–1266.

1970   Pleistocene mammalian local faunas from the Great Plains and Central Lowland provinces of the United States. *In* Pleistocene and Recent Environments of the Central Great Plains, eds. Dort, W. and J.K. Jones. Dept. Geol., Univ. Kansas, Special Publ. No. 3. The University Press of Kansas, Lawrence/Manhattan/ Wichita: 395–433.

HIBBARD, C.W., C.E. RAY, D.E. SAVAGE, D.W. TAYLOR, AND J.E. GUILDAY

1965   Quaternary mammals of North America. *In* The Quaternary of the United States, a review volume for the VII Congress of the International Association for Quaternary Research, eds. Wright, H.E., Jr. and D.G. Frey. Princeton University Press: 509–525.

LINDSAY, E.H.

1972   Small mammal fossils from the Barstow Formation, California. Univ. Calif. Publs. Geol. Sci., 93: 1–104.

MEADE, G.E.

1945   The Blanco fauna. Publ. Bur. Econ. Geol. Univ. Tex., 4401: 509–556.

OSBORN, H.F.

1918   Equidae of the Oligocene, Miocene, and Pliocene of North America, iconographic type revision. Mem. Am. Mus. Nat. Hist., new series, 2: 1–217.

REPENNING, C.A.

1962   The giant ground squirrel *Paenemarmota*. J. Paleont., 36(3): 540–556.

RUSSELL, L.S.

1957   Tertiary plains of Alberta and Saskatchewan. Proc. Geol. Assn. Can., 9: 17–19.

1958   A horse astragalus from the Hand Hills conglomerate of Alberta. Nat. Hist. Pap. Natn. Mus. Can., 1: 1–3.

SCHULTZ, G.E.

1965   Pleistocene vertebrates from the Butler Spring local fauna, Meade County, Kansas. Pap. Mich. Acad. Sci., Arts, Lett., 50: 235–265.

STORER, J.E.

1970a   New rodents and lagomorphs from the Upper Miocene Wood Mountain Formation of southern Saskatchewan. Can. J. Earth Sci., 7(4): 1125–1129.

1970b   The Wood Mountain fauna: an Upper Miocene mammalian assemblage from southern Saskatchewan. PH.D. thesis, University of Toronto, 400 pp.

1972   Mammals of the Hand Hills Formation of Southern Alberta: preliminary faunal list. Blue Jay, 30(2): 119–120.

TYRRELL, J.B.

1887   Report on a part of northern Alberta and portions of adjacent districts of Assiniboia and Saskatchewan. Geol. Nat. Hist. Surv. Can., Ann. Rept., new series, 2: 1E–176E.

WARREN, P.S.

1939   The Flaxville Plain in Alberta. Trans. R. Can. Inst., 22(2): 341–349.

WEBB, S.D.

1969   The Burge and Minnechaduza Clarendonian mammalian faunas of north-central Nebraska. Univ. Calif. Publs. Geol. Sci., 78: 1–191.

New Rhinocerotoids
from the Oligocene of Nebraska

Lloyd G. Tanner
Co-ordinator of Systematic Collections, University of Nebraska
State Museum and Assistant Professor of Geology, Lincoln

Larry D. Martin
Assistant Curator of Vertebrate Paleontology, University of
Kansas Museum of Natural History, and Assistant Professor
of Systematics and Ecology, Lawrence

Abstract
The rhinocerotoid *Penetrigonias hudsoni* gen. et sp. nov. differs from
previously described rhinocerotoids in the unusual crown pattern
developed on upper premolars two and three (P^2–P^3). The hypocones of these teeth are joined to the protocone by a narrow mure
while the metaloph is more closely associated to the protocone than
to the hypocone. A new species of *Hyracodon* is also described.

Introduction

The Chadronian deposits of North America have produced a varied rhinocerotoid
fauna, which was reduced by Orellan times. *Toxotherium* and *Triplopides* are rare
and appear to be exclusively Chadronian forms, while the remains of hyracodontids
are relatively numerous and compose an important part of the Oligocene rhinocerotoid population. Study of the rhinocerotoid remains from the Oligocene of
northwestern Nebraska has produced evidence that a new genus and two new
species occur in these deposits.

Four species of *Hyracodon* have been described from Chadronian sediments:
Hyracodon priscidens, H. selenidens, H. petersoni and *H. browni*. The holotype
of *Hyracodon selenidens* Troxell has usually been assigned to the middle Oligocene

(Troxell, 1921). However, there is a titanothere tooth which should have been collected in association with the holotype since they carry the same number and associated data (YPM 1173, Fig. 1). There is also a maxilla of an old individual of *Hyracodon* and several rami in the University of Nebraska State Museum Study Collections which are from the Chadron Formation and are closely similar to *H. selenidens*. We are not aware of any definite post-Chadronian records of titanotheres and regard the associated titanothere tooth as evidence that *H. selenidens* is a Chadronian form. It has been placed in synonymy with *H. arcidens* by Sinclair (1922) and Wood (1927). Scott (1942) and Radinsky (1967, p. 30) regarded all described species of *Hyracodon* as synonyms of *H. nebraskensis* Leidy. The published descriptions indicate that both *H. arcidens* and *H. nebraskensis* are about 20 per cent larger than *H. selenidens* and both are geologically younger (Orellan?). In our opinion, *H. selenidens* is a valid species. We have not examined the type material of *H. browni*. This form was not illustrated and the complete absence of a metaloph as listed in the diagnosis is not approached by any hyracodontid with which we are familiar. *Hyracodon petersoni* is probably a synonym of *H. priscidens*, but it is about 15 per cent smaller and the name might be retained at the subspecific level. *H. selenidens* seems ancestral to the larger *H. arcidens minimus* Troxell of the middle Oligocene. Examination of several hundred jaws of early to late Oligocene *Hyracodon* in the University of Nebraska State Museum suggests that later species of *Hyracodon* tend to be larger than Chadronian forms and no examples as small as *H. selenidens* were found from Orellan or later sediments.

A new species of Chadronian hyracodontid is described here in which P⁴ is molariform, and is similar in that respect to *Epitriplopus*. The new species differs from that genus in having the protocone on the molars constricted. However, it is not impossible that *Epitriplopus* may have given rise to the Chadronian species through an advanced form such as *E. medius*. The constriction of the protocone found in *Hyracodon* (Radinsky, 1967) seems to be a progressive feature which may have occurred independently in several lineages.

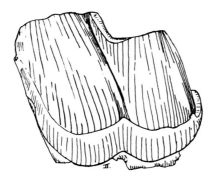

Fig. 1 Titanothere left lower molar, buccal view, YPM 1173. Scale = 20 mm.

Systematics

Class Mammalia
Order Perissodactyla
Superfamily Rhinocerotoidea
Family *Incertae sedis*

Penetrigonias, gen. nov.

Etymology
From Latin *paene* ("near") *Trigonias* (a genus of Oligocene rhinocerotoid with a P^4 similar to that of *Penetrigonias*).

Type Species
Penetrigonias hudsoni sp. nov.

Generic Diagnosis
About 12 per cent larger than *Hyracodon priscidens petersoni* Wood; premolars brachydont; external cingula absent; internal cingula very heavy; metaloph attached to protocone on P^{2-4}; hypocone attached to protocone by a thin spur, and closely adpressed to the internal cingulum on each tooth.

Discussion
The familial affiliations of *Penetrigonias* are at the present uncertain. It is similar to the Hyracodontidae in its small size and general overall resemblance to the Hyrachyidae which we regard as the ancestral stock of the hyracodontids. *Penetrigonias* can be separated from *Hyracodon* by its more brachydont teeth; lack of distinct external cingula on P^{2-4}; heavy internal cingula on these teeth; association of the metaloph with the protocone rather than the hypocone; and the small tear-shaped hypocones which are closely adpressed to the internal cingula. The crown pattern of P^4 resembles that of some species of *Subhyracodon*, while the P^3 of *Trigonias gregoryi* shows similarity to that of *Penetrigonias*. However, none of the Oligocene Rhinocerotidae known to the writers approaches the very small size and combination of tooth characters in *Penetrigonias*; thus close relationship with that family seems unlikely.

In comparison to other Oligocene rhinocerotoids, *Penetrigonias* has retained a very primitive premolar pattern. It is only a more hypsodont state of the pattern found in some species of *Hyrachyus* (Fig. 2c). This pattern is approached in some Eocene amynodonts and, together with the very heavy internal cingulum, might tend to support a tentative assignment to that family. However, P^2 in *Penetrigonias* is less reduced than P^2 of even the Eocene amynodonts and has a prominent mesial interstitial wear facet so that P^1 must have been well-developed. *Penetrigonias* may also be separated from the described species of *Metamynodon*, *Megalamynodon*, *Amynodontopsis* and *Amynodon* by its small size and tear-shaped hypocones. The Asiatic Eocene amynodont *Caenolophus* differs from *Penetrigonias* in that: cingula are not as well developed on the premolars; hypocones are not developed on P^{3-4}; and the medial valley of P_4 is open. It is apparently part of a middle to late Eocene radiation that led to a group of miniature rhinocerotoids with tooth characters resembling those of the amynodonts. *Toxotherium*, which is known from the lower Oligocene of Wyoming, Texas, and Saskatchewan, may be a related genus. The upper dentition of *Toxotherium* is presently unknown. *Penetrigonias hudsoni* is larger than any of the described species of *Toxotherium*, but additional specimens are required to establish the degree of relationship between the two forms. Skinner and Gooris (1966, p. 10) point out that the lower dentitions of *Toxotherium* lack well-developed cingula and that it is unlikely that they would go with an upper dentition that has strong cingula. This might also serve to exclude *Penetrigonias* from *Toxotherium*.

***Penetrigonias hudsoni* sp. nov.**

Fig. 2A, B

Etymology
Named in honour of the late William F. Hudson of Crawford, Nebraska.

Holotype
Left P^{2-4}, UNSM. 62049 (Fig. 2, A-B).

Hypodigm
Type only.

Type Locality
UNSM. Coll. Loc. SX–41 (Sec. 5, T. 33 N., R. 53 W.), located in the central portion of the Chadron Flats north of Toadstool State Park, approximately 26 mi (= 41 km) northwest of Crawford, Sioux County, Nebraska.

Stratigraphic Occurrence
Oligocene, White River Group, upper part of the Chadron Formation.

Specific Diagnosis
Same as for genus.

Description
A small rhinocerotoid with brachydont teeth; P^{2-4} lacking external cingula, but with strong internal cingula; P^2 square with protoloph confluent with a large protocone; protocone with a thin connection to the hypocone; hypocone isolated when unworn and abutting on internal cingulum in P^{2-4}; P^3 squared; metaloph connected to protocone by a thin mure; hypocone joined to protocone by a narrow connection; P^4 subtriangular; metaloph confluent with protocone; hypocone present as a small spur from protocone; parastyles and postfossettes present on P^{2-4}; ectolophs of teeth with "tartar" deposits.

Discussion

The two hyracodontids closest both geographically and temporally to *P. hudsoni* are *Hyracodon priscidens petersoni* and *H. selenidens*. Comparison of the holotypes of these two forms (CM 3572 and YPM 1173 respectively) indicate that they both differ from *Penetrigonias hudsoni* in that: they are smaller; they are more hypsodont; their metalophs are more closely associated with the hypocone than with the protocone; their protocones are not isolated cusps before wear; and they have external cingula and lighter internal cingula.

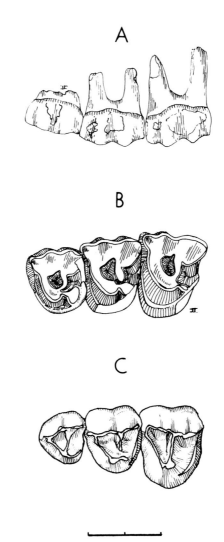

Fig. 2 A–B *Penetrigonias hudsoni*, gen. et sp. nov., holotype, UNSM 62049, left P2-4. Scale = 20 mm.
   A. Buccal view.
   B. Crown view.
   C. *Hyrachyus* sp., UNSM 62099, left P2−4, crown view.

Family Hyracodontidae

Hyracodon doddi, sp. nov.
Fig. 3

Etymology
Named in honour of the late Howard Dodd of Crawford, Nebraska.

Holotype
A right maxilla with P^4-M^2, UNSM. 11066 (Fig. 3A, B).

Hypodigm
Type and one referred specimen, UNSM. 62072 (Fig. 3C).

Type Locality
UNSM Coll. Loc. sx-29, Twin Buttes, 7¾ mi (= 12.5 km) north, 5½ mi (= 9 km) west of Crawford, Sioux County, Nebraska.

Stratigraphic Occurrence
Oligocene, White River Group, Chadron Formation, 18 ft (5.5 m) below the Upper Purplish White Layer (Schultz and Stout, 1955).

Specific Diagnosis
About 13 per cent larger than *Hyracodon priscidens petersoni* Wood and *H. selenidens* Troxell; external cingula on P^4-M^2 restricted to posterior parts of teeth; P^4 molariform with medial valley opening medially (not posteriorly as in *H. priscidens*).

Description
Maxilla slightly smaller than in most *Hyracodon nebraskensis* Leidy; P^4 with metaloph and protoloph straight as in molars, leaving the medial valley open; small antecrochet on protoloph; internal cingulum about as in *H. priscidens petersoni*; protoloph longer than metaloph on P^4 making tooth subtriangular; protocones on P^4-M^2 constricted; cristae, parastyles, postfossettes and antecrochets present on P^4-M^2.

Discussion
Hyracodon doddi is unusual for a Chadronian hyracodontid in having a molariform P^4, a condition that is absent in about 50 maxillae of other species of *Hyracodon* from late Orellan or older deposits examined by us. *Hyracodon doddi* was compapred with referred material of *H. apertus* Sinclair and the holotype of *H. leidyanus* Troxell (YPM 11169) both of which have molariform P^4's. However the younger species are about 20 per cent larger than *H. doddi*. The holotype of *H. arcidens minimus* Troxell (YPM 1174) and *H. selenidens* Troxell have the medial valley of P^4 closed by the protoloph and *H. arcidens minimus* is larger than *H. doddi*. The holotype of *H. priscidens petersoni* Wood (CM 3572) differs from *H. doddi* in having the medial valley opening to the rear on P^4, the metaloph shorter and the protoloph confluent with the hypocone. *Hyracodon doddi* resem-

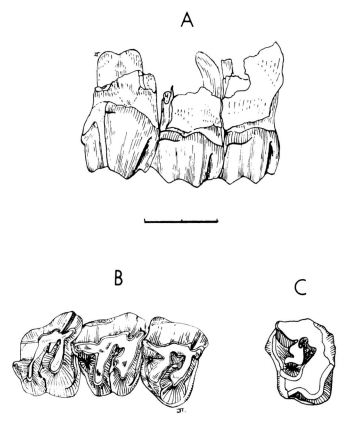

Fig. 3 *Hyracodon doddi*, sp. nov., holotype, UNSM 11066, right P⁴–M².
   A. Buccal view.
   B. Crown view. Scale = 20 mm.
   C. Referred specimen, UNSM 62072, crown view slightly enlarged.

bles closely the late Eocene *Epitriplopus medius* Peterson from the Duchesne Formation of Montana. It differs in having a constricted protocone on the molars.
Epitriplopus medius also lacks the prominent external cingulum on P⁴ and is much
smaller.

In 1964, a specimen was collected from the Chadron Formation at UNSM Coll.
Loc. DW-107 a few miles east of the type locality of *H. doddi*. The size and proportions of this tooth, UNSM 62072 (Fig. 3C), suggest that it may be a P³ of *H.
doddi*. The ectoloph is broken off the tooth, as is a portion of the hypocone. The
protoloph is large and continuous with the hypocone. The protocone and hypocone
are connected by a thick mure. The metaloph is short so that the medial valley
opens posteriorly to the rear, but additional wear would seal off the medial valley
and isolate a lake just posterior and lingual to the medial valley. The metaloph is
posteriorly directed and slightly folded on itself; it bears a labially directed crochet
which isolates a small lake posterior and labial to the medial valley. A postfossette

and internal cingulum are present. The tooth is brachydont, and from a mature but not old individual.

This tooth is smaller and more complicated than that of any of the caenopine rhinocerotoids known to the writers, but it does resemble P³'s of a few species, notably *Trigonias hypostylus*, although that species is much larger.

For tooth measurements of *Penetrigonias hudsoni* and *Hyracodon doddi*, see Table 1.

Table 1. Maximum measurements of tooth diameters (mm) in *Penetrigonias hudsoni* and *Hyracodon doddi*.

Species	Width	Length
Penetrigonias		
P²	17.4	13.2
P³	18.8	16.0
P⁴	22.5	15.2
Hyracodon		
M¹	22.1	15.3
M²	20.5	16.4
M³	22.6	broken

Summary and Conclusions

The taxonomic position of *Penetrigonias* is uncertain and the placement of *Hyracodon doddi* in *Hyracodon* is largely a matter of convenience and may not correctly reflect its relationships. The hyracodonts seem to be a primitive stem-group of rhinocerotoids which have become a depository for primitive forms or, as in the present case, forms known from fragmentary materials. One possible interpretation of the phylogenetic relationships of these forms is shown in Figure 4. If this arrangement is correct, *Penetrigonias* and *Toxotherium* should be placed in a separate family, and *Hyracodon doddi* may belong in *Epitriplopus*. However, until better materials are available it seems useful to retain the more conservative classification in the text of this paper.

Penetrigonias, Toxotherium, Triplopides and *Hyracodon* represent a continuance of a middle and late Eocene radiation of diminutive browsing rhinocerotoids, which were in competition with the early horses. This competition continued unsuccessfully for *Hyracodon* during the Oligocene.

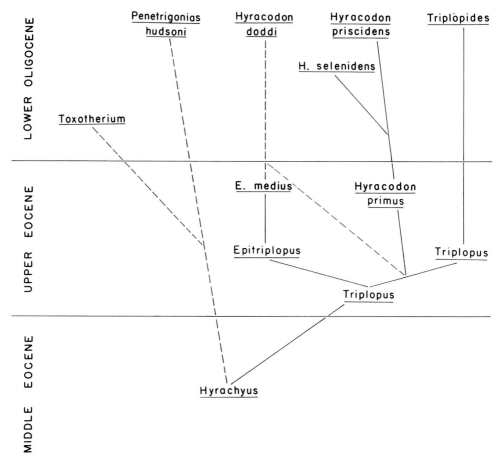

Fig. 4 Chart showing phylogenetic relationships of the Upper Eocene and Lower Oligocene small rhinocerotoids in North America (partly modified after Radinsky, 1967).

Acknowledgments

The authors are grateful to C.B. Schultz, Malcolm C. McKenna, and Leonard B. Radinsky for helpful comments, and to Craig C. Black for the loan of the holotypes of *Epitriplopus medius, Hyracodon petersoni* and *H. primus* from the Carnegie Museum at Pittsburgh. We are also grateful to Elwyn L. Simons for the loan of the holotypes of *H. selenidens, H. arcidens minimus*, and *H. leidyanus* from the Yale Peabody Museum and to Loris S. Russell for a cast of the type of *H. priscidens*. Mary Cutler did the typing and the chart. Jerry Tanner prepared the other illustrations.

Literature Cited

RADINSKY, L.B.
  1967   A review of the rhinocerotoid family Hyracodontidae (Perissodactyla). Bull. Amer. Mus. Nat. Hist., 136(1): 1–46, figs. 1–25, 1 pl., 6 tables.

SCHULTZ, C.B. AND T.M. STOUT
  1955   Classification of Oligocene sediments in Nebraska. Bull. Univ. Nebr. State Mus., 4(2): 17–52, figs. 1–12.

SCOTT, W.B.
  1941   Perissodactyla. *In* Scott, W.B., and G. L. Jepson, The mammalian fauna of the White River Oligocene. Trans. Amer. Phil. Soc., n.s., 28(5): 747–980.

SINCLAIR, W.J.
  1922   Hyracodons from the Big Badlands of South Dakota. Proc. Amer. Phil. Soc., 61(E): 64–79, figs. 1–8, 1 chart.

SKINNER, S.M. AND R.J. GOORIS
  1966   A note on *Toxotherium* (Mammalia, Rhinocerotoidea) from Natrona County, Wyoming. Amer. Mus. Novitates, 2261: 1–12, figs. 1–5, 1 table.

TROXELL, E.L.
  1921   New species of *Hyracodon*. Amer. Jour. Sci., 202(ser. 5, no. 2, art. 3): 34–39.

WOOD, H.E., II
  1927   Some early Tertiary rhinoceroses and hyracodonts. Bull. Amer. Paleont., 13(50): 1–104, 7 pl., 5 tables.

Observations on the Pygmy Mammoths of the Channel Islands, California

D. A. Hooijer
Rijksmuseum van Natuurlijke Historie, Leiden,
The Netherlands

Abstract

Data on *Mammuthus exilis* (Stock and Furlong) from San Miguel Island, California, are presented and compared with those on *Mammuthus imperator* (Leidy) and the classic Mediterranean and recently discovered Indonesian pygmy proboscideans. Of the Mediterranean species, only *Elephas mnaidriensis* Leith Adams is larger than, and all others are smaller than, *M. exilis*; all three Indonesian pygmy species are smaller.

Introduction

In a survey of the world's Pleistocene pygmy proboscideans (Hooijer, 1967), I included *Mammuthus exilis* (Stock and Furlong, 1928) from the Channel Islands off the coast of southern California merely on the basis of the available literature. In the spring of 1970, while serving as a Regent's Professor at the University of California, Irvine, I became acquainted with Mr. and Mrs. James B. Stoddard of Corona del Mar, California, who invited me to join them on a boat trip to San Miguel Island to observe the fossiliferous beds there. Permission for this trip was granted by Mr. Donald M. Robinson, Superintendent, United States Department of the Interior, National Park Service, Channel Islands National Monument. The trip was made aboard the vessel *Florence A* with Mr. Charles A. Repenning, United States Geological Survey, Menlo Park, California. I made on-site observations in various parts of San Miguel Island during the period of May 22 to 27, 1970. Afterwards, in the Los Angeles County Museum of Natural History (LACM) and in the Earth Sciences Building, University of California, Berkeley, I examined specimens of *Mammuthus exilis* from Santa Rosa Island as well as of *Mammuthus imperator* (Leidy), the large continental Pleistocene mammoth from which the pygmy Channel Islands species is thought to have been derived. My observations and comparisons are recorded below.

Observations and Discussion

In Table 1 are given the data on three M_3's of *Mammuthus exilis*, the Santa Rosa specimens being in the Los Angeles County Museum and the San Miguel M_3 in Dr. Johnson's collection.

The small number of plates (16 as opposed to 16+ and 18) could mean that the San Miguel M_3 was sampled from a somewhat more primitive population than those from Santa Rosa. However, low numbers such as these may be accounted for by individual variation within a single and more advanced species. A variability of two plates and within two units in laminar frequency is common in series of molars of better known populations of other advanced fossil proboscidean species.

The virtual identity of the San Miguel *Mammuthus* species with *Mammuthus exilis* from Santa Rosa is substantiated by the distal part of a left M^3 observed by us in a gully on San Miguel (Table 2). The San Miguel M^3, with nine plates, of which two are worn, differs from those of Santa Rosa neither in width nor in laminar frequency, while the crown height, which is twice the molar width, is similar to that in incomplete specimens of M^3 from Santa Rosa seen by me in the Los Angeles County Museum.

Upper and lower third molars of *Mammuthus imperator*, the large Imperial Mammoth of North America, show the same range in the number of ridges and similar laminar frequencies to those of the insular type (Table 3).

Though properly fitting into the category of "pygmy elephants," *Mammuthus exilis* is nowhere near the smallest of the world's pygmy proboscideans. *Elephas* (*Archidiskodon*) *celebensis* (Hooijer) has an M_3 with 11 plates in the mesiodistal length of 164 mm (Hooijer, 1954), and thus a laminar frequency of 6.6. Moreover, its congener in the Pleistocene of Celebes, *Stegodon sompoensis* Hooijer, has an M_3 with 10 plates in a mesiodistal length of 135 mm (Hooijer, 1972b), to give a laminar frequency of 7.4. Indistinguishable from the Celebes stegodont M_3's are those of *Stegodon timorensis* Sartono from the Pleistocene of Timor, which vary in length from 118 to *ca.* 145 mm and have 10 plates (Hooijer, 1969 and 1972a), and thus have laminar frequencies from 8.5 to 6.9.

Like the molars from the Channel Islands, these from the dwarf Indonesian elephants are half-scale miniatures of the homologues in their alleged parental species, *Elephas planifrons* Falconer and Cautley and *Stegodon trigonocephalus* Martin, but are proportionally slightly longer mesiodistally and have somewhat higher crowns. Whether there is a difference in relative crown height between *Mammuthus imperator* and *M. exilis* remains to be established. Of the three classic Mediterranean pygmy elephants, *Elephas mnaidriensis* Leith Adams, *E. melitensis* Falconer, and *E. falconeri* Busk, the lengths of the M_3's are less than those in *Mammuthus exilis* in two of the species. Whole M_3's with 14 plates from

Table 1. Measurements of lower third molars (M₃'s) of *Mammuthus exilis* (Stock and Furlong) in mm.

Specimen	No. of plates	Mesiodistal length	Buccolingual width	Crown-root height	Laminar frequency per 100 mm mesiodistal length
Santa Rosa (LACM)	18	230	ca. 60	—	7.8
Santa Rosa (LACM)	16+	190+	60	ca. 90	8.4
San Miguel (Johnson Coll.)	16	200	69	83	8.0

Table 2. Measurements of upper third molars (M³'s) of *Mammuthus exilis* (Stock and Furlong) in mm.

Specimen	No. of plates	Mesiodistal length	Buccolingual width	Crown-root height	Laminar frequency per 100 mm mesiodistal length
Santa Rosa (LACM)	14+	172	70	115+	8.2
Santa Rosa (LACM)	18	ca. 190	62	105+	9.5
San Miguel (field obs.)	9+	—	68	140	9 est.

Table 3. Measurements of upper and lower third molars of *Mammuthus imperator* (Leidy) in mm.

Molar	No. of plates	Mesiodistal length	Buccolingual width	Laminar frequency per 100 mm of mesiodistal length
M³	17–19	200–256	106–120	8.5–7.4
M₃	16–17	228–324	96–120	7.0–5.3

Elephas mnaidriensis vary in length from 192 to 220 mm (Piccoli *et al.*, 1970), while in *E. melitensis* and *E. falconeri* mesiodistal lengths of whole M_3's may be as little as 153 mm and 143 mm, respectively (Vaufrey, 1929).

The specimens observed by Mr. Repenning and myself on San Miguel Island occur in a chocolate-brown, partly lateritic siltstone, just above cross-bedded sands containing travertine deposits and below white dune sands with caliche at and near the base. Some specimens we saw exposed to such an extent that they could easily weather away and thus be lost to science unless they are saved and put in a museum. Unfortunately these sediments are very local, and have not been found across the Channel in continental deposits. However, I know of no other small molars such as those from the Channel Islands found elsewhere in the North American continent. Elephant material also occurs at the Running Springs Locality in association with archaeological material (Donald Johnson, *pers. comm.*). This is probably more recent than the specimens observed by us.

Stock (1936, 1943) notes the variation in adult size of *Mammuthus exilis* from Santa Rosa, which ranged from six to nine feet (1.8–2.7 m) in shoulder height. When the Santa Rosa Peninsula became inundated to form the islands, there was probably a progressive diminution in size in the marooned populations. The name *Mammuthus exilis* must, of course, apply to the type specimen, which is the skull and mandible from Santa Rosa designated as the type by Stock and Furlong (1928) and figured by Stock (1943). When it was described, the specimen (No. 14) was part of the Vertebrate Paleontology Collection in the California Institute of Technology, but it is now in the collection of the Los Angeles County Museum. Further collecting at specific levels, with $_{14}$C-controlled dating of the levels on all three islands, may demonstrate progressive size diminution which paralleled the evolutionary drift of the elephants exiled on the Mediterranean islands. The extinction of the dwarf mammoths on the Californian Channel Islands was certainly accomplished mainly by man, as suggested by the charred bones reported by Orr (1960). In 1970 C.A. Repenning and I found no large molars, but a large astragalus and a large calcaneum on San Miguel that are not or hardly smaller than their homologues in *Mammuthus imperator* (Table 4). The *M. imperator* specimens are from Site 5, Tule Springs, Clark Co., Nevada, of Irvingtonian age (*ca.* 1.5 m.y.), and were measured by me at the Museum of Paleontology, University of California, Berkeley. We also found a rather small elephantine scapula on San Miguel for which the dimensions of the glenoid fossa are 120 by 70 mm, compared with 200 by 100 in a specimen of *Mammuthus imperator* in the Los Angeles County Museum.

Table 4. Measurements of tarsals of *Mammuthus* cf. *exilis* and *M. imperator* in mm.

Tarsal	*M.* cf. *exilis* from San Miguel, California	*M. imperator* from Tule Springs, Nevada	
Astragalus			
greatest width	ca. 140	140	150
greatest height	ca. 110	110	120
lateral height	90	85	100
anteroposterior diameter from trochlea to medial calcaneal facet	—	80	ca. 90
Calcaneum			
greatest diameter over fibular and medial astragalar facets	ca. 120	135	145
greatest height	ca. 140	185	200
Anteroposterior diameter of corpus above sustentaculum	70	85	—

The pygmy elephants from both the Mediterranean and Indonesian islands tend to be associated with giant rodents, i.e. *Leithia melitensis* (Adams.) from Malta and *Papagomys* sp. from Flores. This is also so in the Channel Islands of California: Wilson (1936) described a distinct *Peromyscus, P. nesodytes* Wilson, a mandibular ramus of which was discovered among the mammoth remains from Santa Rosa Island, and which is characterized by its very large size. Rodents grow to a large size on islands only when carnivores are absent, as far as can be established from observations in other parts of the world.

A small fox is still extant on the Channel Islands, as I stumbled upon a den with one cub in one of the gullies on San Miguel that we were exploring. Stock (1936, 1943) figures a series of pelts taken from the mainland, Santa Cruz, Santa Rosa, and San Miguel, showing the small size of the island forms as compared with those from the mainland. Whether there were large Pleistocene foxes on the islands is uncertain, as no fossil remains have so far been found. Stock notes that it is possible that the small size of the insular foxes was a subspecific character already exhibited by this population prior to their isolation.

It is clear that much collecting on the islands should be done to enable us to enlarge upon the above observations, and I hope that this may be done with the permission of enlightened authorities before it is too late.

Acknowledgments

It is a great pleasure to thank Mr. and Mrs. James B. Stoddard, Mr. Charles A. Repenning, Dr. Theodore Downs, Mr. David Fortsch, and Dr. Donald E. Savage who made this work possible under such pleasant circumstances. The findings have been reported to Mr. Donald M. Robinson as well as to Dr. Donald L. Johnson, Department of Geography, Southern Illinois University, Carbondale, Illinois, who kindly made available a cast in his possession of an entire left lower last molar from San Miguel, the only specimen of the San Miguel elephant known to me before the trip to the island in 1970.

Literature Cited

HOOIJER, D.A.

1954   Pleistocene vertebrates from Celebes. XI. Molars and a tusked mandible of *Archidiskodon celebensis* Hooijer. Zool. Med. Mus. Leiden, 33(15): 103–120.

1967   Indo-Australian insular elephants. Genetica, 38: 143–162.

1969   The *Stegodon* from Timor. Proc. Kon. Nederl. Akad. Wetens. Amsterdam, ser. B, 72: 203–210.

1972a   *Stegodon trigonocephalus florensis* Hooijer and *Stegodon timorensis* Sartono from the Pleistocene of Flores and Timor. Proc. Kon. Nederl. Akad. Wetens. Amsterdam, ser. B, 75: 12–33.

1972b   Pleistocene vertebrates from Celebes. XIV. Additions to the *Archidiskodon-Celebochoerus* fauna. Zool. Med. Mus. Leiden, 46(1): 1–16.

ORR, P.C.

1960   Radiocarbon dates from Santa Rosa Island. II. Santa Barbara Mus. Nat. Hist., Dept. Anthropol., Bull. no. 3: 1–5.

PICCOLI, G., O.L. FORMENTIN, G.W. DEL PUP AND M.E.Z. VISENTIN

1970   Studi su resti di crani di *Elephas mnaidriensis* Adams del Pleistocene di Sicilia. Mem. Ist. Geol. Min. Univ. Padova, 27: 1–33.

STOCK, C.

1936   Ice Age elephants of the Channel Islands. Westways, June 1936: 14–15.

1943   Foxes and elephants of the Channel Islands. New discoveries on the Channel Islands. Los Angeles Co. Mus. Quart., 3: 6–8.

STOCK, C. AND E.L. FURLONG

1928   The Pleistocene elephants of Santa Rosa Island, California. Science, 68: 140–141.

VAUFREY, R.

1929   Les éléphants nains des îles méditerranéenes et la question des isthmes pléistocènes. Arch. Inst. Pal. Humaine, Mém., 6: 1–220.

WILSON, R.W.

1936   A new Pleistocene deer-mouse from Santa Rosa Island, California. J. Mammal., 17: 408–410.

Notes on American Pleistocene Tapirs

Ernest L. Lundelius, Jr.
Department of Geological Sciences, The University of Texas
at Austin

Bob H. Slaughter
Shuler Museum of Paleontology, Southern Methodist
University, Dallas

Abstract

A skull of *Tapirus veroensis* from the Livingston Dam on the Trinity River, Texas, is dated at 21,000 y BP. Comparison of the skull with others of Pleistocene tapirs suggests that *T. haysii* and *T. tennesseae* are inadequately based, that *T. californicus* and *T. merriami* require additional and better material for validation, and that *T. copei* is probably a valid species. *T. excelsus* is a growth stage of *T. veroensis*, and when larger samples of Pleistocene tapirs become available from the eastern United States, the larger individuals may become known as *T. veroensis excelsus*.

Introduction

In 1965, a party from the Shuler Museum of Paleontology of Southern Methodist University mapped Quaternary deposits along a section of the Trinity River basin in Polk and San Jacinto Counties, Texas. This area was to be flooded by waters impounded by the Livingston Dam then under construction. In the dam-excavation and immediately downstream were small exposures of a gravel bed four feet (1.2 m) thick, the base of which was almost at stream grade. This is designated the Wayne Baker Farm locality. The gravel bed contained numerous logs and limbs (mostly oak) and the upper half contained ferruginous cement in many places. One stump was rooted in the underlying Miocene clay and covered by the Pleistocene gravel. A sample of the wood was processed by The Radiocarbon-Dating Laboratory of the Shell Research Center, Houston, Texas, through the good offices of Mr. Owen Baker. The radiometric analysis gives an age of 21,590 ± 570 BP (Slaughter, 1965). Associated with the logs were bones of several animals. These fossils were collected by personnel of Southern Methodist University, Rice University, and the Houston Museum of Natural Science. Two especially well-preserved skulls were recovered from this locality by Dr. Thomas Pulley, of the Houston Museum. One of these is from *Megalonyx jeffersonii* and the other from a tapir. Good skulls of these animals are rare in the United States, especially ones that can be accurately dated. The tapir skull (Fig. 3) is discussed together with the validity of most of the proposed specific names.

Before Sellards' (1918) description of a fine skull (Fig. 1E) and associated lower jaw fragment from Florida, most American Pleistocene tapir material was referred either to *Tapirus haysii* or to the living Central and South American form, *T. terrestris*. These references were usually made on the basis of size alone; the smaller specimens were considered *T. terrestris*, and the larger ones, *T. haysii*. Among the fossil material referred to *T. haysii* is a collection from the Port Kennedy Cave, Pennsylvania. When Cope (1899) described this collection he too referred the fossil to *T. haysii*, noting its large size. As the Port Kennedy Cave collection includes both upper and lower jaws, subsequent comparisons were made with Cope's referred material rather than the holotype, a single tooth.

Although the Florida skull fell well within the size-range for the living form, *T. terrestris*, Sellards (1918) adequately demonstrated that the fossil represented a species other than any living and proposed the name *T. veroensis*. This name then replaced *T. terrestris* as the usual reference to American Pleistocene tapirs that were smaller than the tapir represented at Port Kennedy.

Simpson (1945) applied a statistical approach to a large sample of *T. terrestris* and a number of specimens referred to *T. veroensis* from Seminole Field, Florida, and re-examined the other described fossil tapirs from the United States. He demonstrated that the type specimen of *T. haysii* was probably within the size range of *T. veroensis* but approached the lower limit of the size of the Port Kennedy tapir. The holotype of *T. haysii* is not only intermediate but there is considerable doubt as to the position from which the tooth originated. Simpson, therefore, considered the holotype to be at least unidentified if not unidentifiable. For that reason he proposed a new name for the Port Kennedy material, *Tapirus copei*.

Simpson also concluded that *T. tennesseae* Hay was based on a collection that included isolated teeth of more than one individual and that the taxon was not identifiable. He considered it a possible synonym of *T. veroensis* but intermediate and "properly ignored".

In the same paper (1945) Simpson pointed out that most of the characters of tapir dentitions that have been utilized in the separation of species are suspect. Our observations have tended to substantiate his view. Such features as the presence, absence, or strength of basal tubercles between the lophs (both lingual and labial) of upper and lower teeth are extremely variable in all American species of *Tapirus*. If large samples were known from each locality, statistical analysis might prove helpful at a population or subspecific level; but since there are only two collections of adequate size (from Seminole Field and the Melbourne Formation), this aspect is not considered here.

The degree of molarization of P^1 and/or P^2 appears to be the most promising feature of the dentition to differentiate geographic populations but even this must be treated with caution. The weakest development of the protocone on P^1 among American Recent and Pleistocene forms is that of *T. pinchaque* (=*T. roulini*). Although there are not many specimens available of this relatively rare species, the condition is consistent in the three specimens we have examined. The P^1 of *T. copei* appears to have the greatest development of the protocone. Cope (1899) and later Simpson (1945) described the slightly more forward position of the protocone and the ridge joining it to the ectoloph at about its mid-length. This forward position creates a hypoconal basin that is not well developed in other American fossil tapirs we have seen.

   T. californicus and *T. merriami* were based on isolated teeth but were considered by Simpson as possibly valid forms. He did point out, however, that both are within the probable size range of *T. copei* and that better material from California is needed to clarify their position.

Description

Texas Skull

The 21,000-year-old skull from the Livingston Dam is complete except for the zygomatic arches, the right P^1 and M^3, the nasals, the premaxillae, and the anterior portions of the maxillae. The specimen is comparable to the holotype of *T. veronensis* except for its slightly larger size. The complete dentition is erupted and in use. Dental wear indicates a similar age for the Texas and Florida individuals although the one from Florida may be slightly older as shown by the greater fusion of the sutures.

   Considering Sellard's (1918) excellent description of the skull of *T. veroensis* and Simpson's (1945) additional comparison of a cast of that specimen with other species of *Tapirus*, we think it expedient to comment on the new specimen and additional comparative material and on the 13 skull characters considered important by Sellards and Simpson.

   Simpson (1945) gives the characters:
   (1) Dorsal contour.
   (2) Sagittal crest of sagittal table.
   (3) Dorso-occipital triangular table.
   (4) Relative width and inflation of the frontals.

   These four characters are all controlled by the character of the sagittal crest (or table). This will be discussed in some detail below.

Fig. 1 A. Dorsal view of cranium of holotype of *T. excelsus* Simpson (1945), M^1 erupted.
   B. Dorsal view of cranium referred to *T. veroensis* (FSM 14064). M^1 and M^2 erupted; P^4 erupting.
   C. Dorsal view of immature cranium of *T. pinchaque*. M^1 and M^2 erupted.
   D. Dorsal view of cranium referred to *T. excelsus* by Oesch (1967) (CM 159). M^1, M^2 and P^4 erupted.
   E. Dorsal view of cranium of holotype of *T. veroensis* (FGS V277). All permanent teeth erupted and in use.
   F. Dorsal view of cranium of *T. pinchaque* (SMP-SMU 62723). All permanent teeth erupted except M^3.

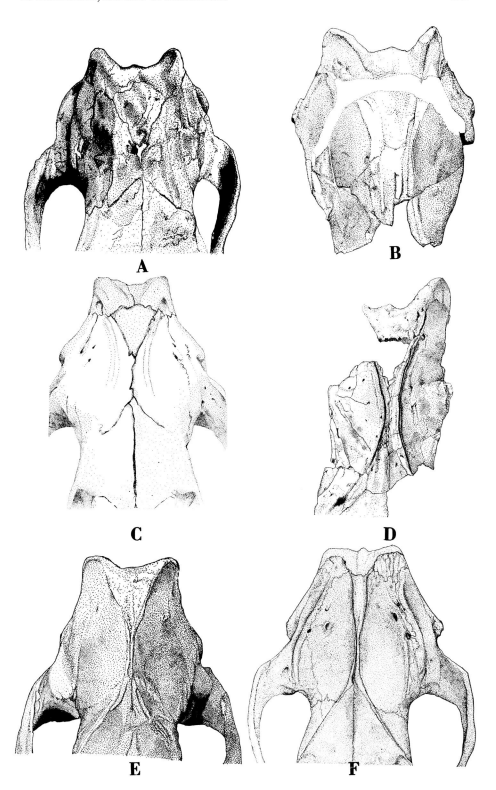

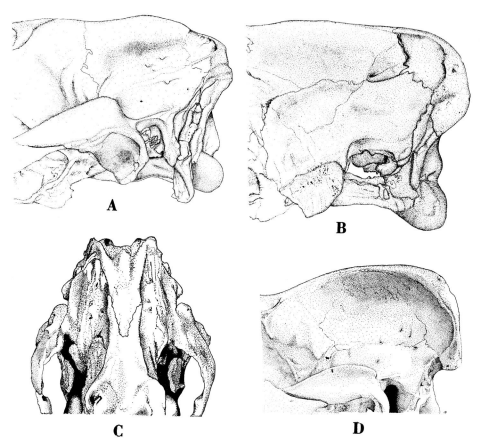

Fig. 2 A. Lateral view of cranium of adult *T. pinchaque* (SMP-SMU 62723).
        B. Lateral view of cranium of Texas Skull referred to *T. veroensis* (HMNS 160).
        C. Dorsal view of cranium of *T. bairdii* (AMNH 130104).
        D. Lateral view of cranium of *T. terrestris* (AMNH 78519).

(5) The spiral grooves that circle the postero-labial corner of the nasals are very broad and shallow in the Texas fossil. In this feature, it resembles the Florida fossil skull more closely than any living species.

(6) The lateral grooves leading to the spiral grooves are also broad and shallow. Simpson notes that this is the case in *T. indicus* but unlike the deeper and narrower lateral grooves of *T. pinchaque*, *T. terrestris*, and *T. bairdi*. We feel that this minor similarity with *T. indicus* is less significant than the sagittal region in suggesting relationships and probably is a mere reflection of the cranial length-width ratios and other rather general aspects of skull shape.

(7) There is a fairly well-developed lacrimal pit in both the Texas and Florida fossils. Simpson notes, however, that there always seems to be some suggestion of such a pit in tapirs and does not feel that this was particularly significant.

(8) The character of the ascending process of the premaxillae; this area is not well preserved in the new specimen.

(9) Both *T. veroensis* and the Texas fossil have maxillae with sharp, somewhat

recurved dorso-medial edges. These edges are rounded in *T. terrestris* and *T. pinchaque*. The condition in the fossil is most closely matched by *T. indicus*.

(10) The lateral convexity of the palate in the holotype of *T. veroensis* is greater than in any of the living forms. The anterior end of the palate of the Texas skull is indeed more convex than any Recent specimens we have examined but, posterior to P^2, the convexity seems to be about the same as that of *T. pinchaque*.

(11) There is a rugosity at the posterior end of the palate just below the vomer. This was listed by Simpson as present in *T. veroensis* and *T. indicus*, absent in *T. pinchaque*, and variable in *T. terrestris*. Our observations of additional specimens of these species substantiate his conclusions.

(12) The lamboidal crests flare outward and backward and are rather strong in *T. veroensis*, *T. pinchaque*, and *T. indicus*. The crests are somewhat weaker in *T. terrestris* and *T. bairdi*.

In the same comprehensive paper, Simpson (1945) proposed a new species, *Tapirus excelsus*, based on a partial skeleton of an immature individual from a fissure-fill in Missouri. The only permanent teeth represented in the holotype are unerupted P^1 and M^1. The width of each of these teeth is just within the upper size limit of the observed range for *T. veroensis*, but the antero-posterior diameters are beyond the range observed in the Florida collection.

A second specimen (Fig. 1D) from another fissure in Missouri was tentatively referred to *T. excelsus* by Oesch (1967). This specimen is also large compared to *T. veroensis*, but the dentition has measurements close to those of the new Texas skull and one from Oklahoma that was referred to *T. haysii* by Stovall and Johnston (1934). Additional tapir material from the Ingleside site in Texas has been referred to *T. excelsus*, on the basis of the size and proportions of the teeth (Lundelius, 1972). Simpson was unable to distinguish the dentition of the holotype of the Missouri species from that of *T. veroensis*, except by its slightly larger size and different length-width ratios of dP_4, but he placed considerable weight on the character of the sagittal region of the skull.

The holotype of *T. excelsus* does not have a medial sagittal crest as seen in the holotype of *T. veroensis*. Rather the junction of the parietals is broad and flat (Fig. 1A). Lateral to this "sagittal table" are two acutely rounded "escarpments". A similar arrangement occurs in the living forms, *T. bairdi* (Fig. 2C) and *T. indicus*, but the "escarpments" are somewhat sharper and more distinct. On the other hand, *T. veroensis*, the new Texas skull, and the living species *T. pinchaque* have the more familiar sagittal crest. In spite of the subdued nature of the "escarpments" on the Missouri specimen, Simpson felt that in this feature at least, *T. excelsus* was most like *T. bairdi* and *T. indicus*.

He was aware that the young of species that have sagittal crests, as adults often have similar sagittal tables (e.g. peccaries, horses, etc.). However, he examined the skull of an immature individual of *T. terrestris* that was of about the same age when it died as the Missouri specimen and found a sagittal crest already clearly in evidence. He therefore reasoned that the degree of immaturity of the Missouri tapir did not have a significant effect on this feature. It is true that *T. terrestris* has a sagittal crest, but it should be pointed out that it is extremely well developed. Indeed, it involves about one half of the surface of the parietals (Fig. 2D). Such a

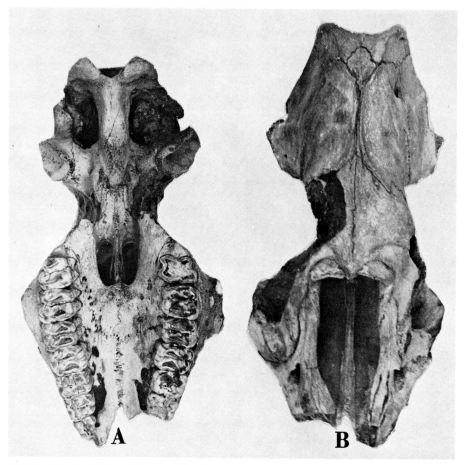

Fig. 3 A. Ventral view of Texas skull referred to *T. veroensis* (HMNS 160).
B. Dorsal view of skull referred to *T. veroensis* (HMNS 160).

tremendous development might be expected to be initiated earlier than in animals with minimal development of the crest.

The saggittal crest of *T. pinchaque* (Fig. 1F) is low and developed to the same degree as it is in *T. veroensis* and the new Texas skull. These crests are formed by the slight upward crimping of the parietals at their junction. The fact that young specimens of *T. pinchaque* were not available to Simpson (1945) explains his use of *T. terrestris* for comparison. Not only have we examined a young *T. pinchaque* skull, but there are now two additional fossil skulls in which this region is preserved. The young *T. pinchaque* skull (Fig. 1C) has a sagittal table with a pair of "escarpments" not unlike those of the holotype of *T. excelsus*. The "table" is narrower than that of Simpson's specimen, but this individual, judging by the eruption of M^2, was slightly older at the time of death. A fossil from the Itchtucknee River, Florida (Fig. 1B; UF 14064) has M^1 and M^2 fully erupted, but P^4 is just coming in. This specimen also has a sagittal table but it is narrower than that of the young *T. pinchaque* skull. The measurements of the teeth of this specimen fall well within the demonstrated size range for *T. veroensis*.

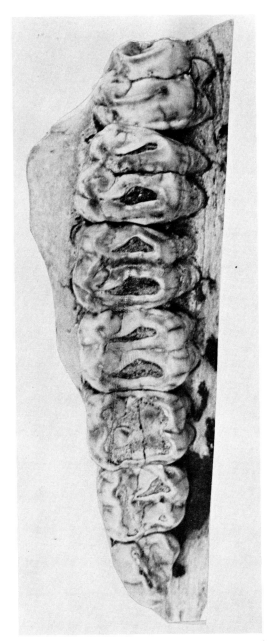

Fig. 4 Occlusal view of left cheek-tooth series of Texas specimen referred to *T. veroensis* (HMNS 160).

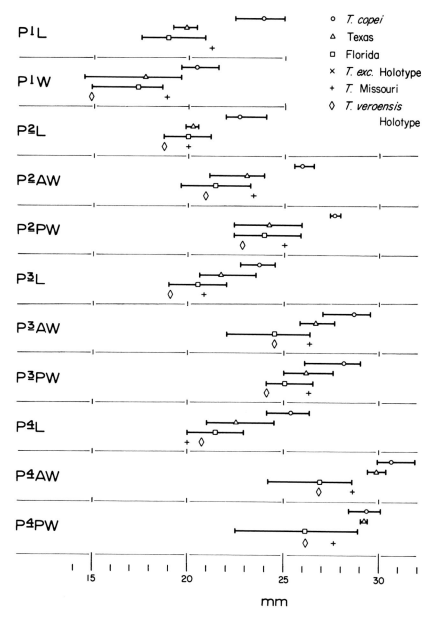

Fig. 5 Bar graphs showing means and observed ranges of measurements of upper premolars of various samples of Pleistocene tapirs. Data for *T. copei* from Simpson (1945). L — mesiodistal length; w — buccolingual width; AW — anterior width; PW — posterior width.

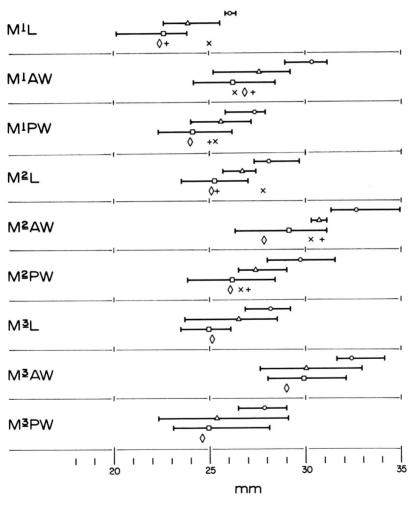

Fig. 6 Bar graphs showing means and observed ranges of measurements of upper molars of various samples of Pleistocene tapirs. Symbols as in Fig. 5. Data for *T. copei* from Simpson (1945). L — mesiodistal length; w — buccolingual width; AW — anterior width; PW — posterior width.

Table 1. Numerical data on upper dentitions of *Tapirus veroensis* from Texas. L = mesiodistal length; W = buccolingual width; AW = anterior width; PW = posterior width.

Tooth	Mean (mm)	n	Observed range (mm)
P^1 L	19.8	4	19.2–20.4
P^1 W	17.7	4	14.5–19.6
P^2 L	20.2	3	19.8–20.5
P^2 AW	23.0	4	21.1–23.9
P^2 PW	24.2	4	22.4–25.9
P^3 L	21.7	5	20.6–23.5
P^3 AW	26.6	5	25.8–27.6
P^3 PW	26.2	5	25.0–27.6
P^4 L	22.5	3	21.0–24.5
P^4 AW	29.8	3	29.4–30.3
P^4 PW	29.2	3	29.1–29.3
M^1 L	23.9	6	22.6–25.6
M^1 AW	27.6	6	25.2–29.2
M^1 PW	25.6	6	24.0–27.2
M^2 L	26.7	4	25.7–27.4
M^2 AW	30.7	3	30.3–31.1
M^2 PW	27.3	4	26.5–29.0
M^3 L	26.5	3	23.7–28.5
M^3 AW	30.0	3	27.6–32.9
M^3 PW	25.27	3	22.3–29.1

Table 2. Statistical data on upper dentitions of Rancho LaBrean *Tapirus veroensis* from various localities in Florida. L = mesiodistal length; W = buccolingual width; AW = anterior width; PW = posterior width.

Tooth	Mean ± S.E.	Standard deviation	Coefficients of variation	n	Observed range (mm)
P^1 L	18.9±0.27	0.94	5.1	12	17.5–20.8
P^1 W	17.3±0.38	1.32	7.8	12	14.9–18.6
P_2 L	19.9±0.29	0.88	4.5	9	18.7–21.1
P^2 AW	21.5±0.34	1.07	5.1	10	19.6–23.2
P_2 PW	23.9±0.30	0.95	4.1	10	22.8–25.8
P_3 L	20.5±0.33	1.03	5.1	10	19.0–22.0
P_3 AW	24.5±0.37	1.16	4.8	10	22.0–26.3
P^3 PW	25.1±0.23	0.74	3.1	10	24.1–26.5
P^4 L	21.5±0.26	0.93	4.4	13	20.0–22.9
P^4 AW	26.9±0.36	1.31	5.0	13	24.2–28.6
P^4 PW	26.2±0.47	1.69	6.5	13	22.5–28.9
M^1 L	22.6±0.26	1.02	4.6	15	20.2–23.8
M^1 AW	26.3±0.27	1.04	4.1	15	24.2–28.4
M^1 PW	24.2±0.29	1.10	4.7	14	22.3–26.2
M^2 L	25.3±0.29	1.18	4.8	17	23.5–27.0
M^2 AW	29.2±0.31	1.29	4.5	17	26.3–31.1
M_2 PW	26.2±0.27	1.12	4.4	17	23.8–28.4
M_3 L	24.9±0.30	0.96	4.0	10	23.5–26.1
M^3 AW	29.8±0.44	1.39	4.8	10	28.0–32.1
M^3 PW	24.9±0.43	1.35	5.5	10	23.1–28.1

Finally, the Missouri specimen described by Oesch (1967) has the complete dentition erupted except for M^3. There are two crests in this specimen (Fig. 1D), and the parietal suture is visible between them. It seems that they would have merged into a single sagittal crest. It should be noted at this point that the only two North American Pleistocene tapir skulls in which temporal muscles met at the midline had the complete dentition erupted and in use. The more open sutures of the new Texas skull, compared with those of the holotype of *T. veroensis*, suggest a slightly younger individual, and we think it noteworthy that there are two distinct crests even though they are tightly abutted. The Florida skull has its cranial sutures fused and the sagittal crest is a single entity.

From this sequence of specimens it is therefore evident that *T. terrestris*, with its exaggerated sagittal crest, initiates its development early and *T. pinchaque* and *T. veroensis* initiate the development of this structure somewhat later in life. Indeed, it would appear that in *T. veroensis* the two crests do not meet until sometime between the eruption of M^2 and M^3 and do not fuse into a single crest until after M^3 has erupted. Another interesting aspect of the two Missouri skulls and the one from Texas is the presence of a wedge formed by one or two interparietals between the posterior margins of the parietals. They are symetrical about the midline and are bordered posteriorly by the supraoccipital. The type of *T. excelsus* has two such bones, the smaller being the more anterior. The space is empty in Oesch's Missouri specimen but it is apparent that there was at least one interparietal in that position. The Texas skull has but a single diamond-shaped interparietal but the sutures are open. Although there is no sutural evidence of interparietals in the holotype of *T. veroensis*, the area is flat and shaped exactly as it is in the Texas skull and, considering the degree of other sutural closure, it is certain that there was at least one distinct interparietal early in its life. If such a bone, or bones, were present in *T. pinchaque*, they were much smaller. One specimen available to us has the dentition fully erupted, except M^3 (younger than the Texas fossil); there is no sutural evidence of an interparietal, nor is there room for more than a tiny bone in this position.

In the course of this study we compared the sizes of the teeth of a number of samples of Pleistocene tapirs, for which the statistical summaries are given in Tables 1–3. Although the mean values of the Florida sample are smaller than those of the samples from Texas, Oklahoma, and Missouri, there are considerable overlaps in the observed ranges. This is consistent with the interpretation that they represent geographic variants of a single widespread species.

The mean values of all the samples here placed in *T. veroensis* are smaller than those of *T. copei* but with overlaps in the observed ranges of many measurements (Tables 1–3, Figs. 5, 6). The major differences in dental dimensions between *T. copei* and *T. veroensis* are in the premolars and may be a reflection of the increased molarization of P^{1-2} in the former species.

The coefficients of variation for the dental dimensions are within the range to be expected for mammals (Simpson, Roe, and Lewontin, 1960, p. 91).

Scatter diagrams of a number of pairs of measurements have been prepared and in general support the conclusion reached from the statistical summaries in showing that the *veroensis-excelsus* samples constitute a single coherent group that is separable from that of *T. copei*. The scatter diagram of length vs. anterior

Table 3. Numerical data on lower dentitions of *Tapirus veroensis* from various localities in Florida. L = mesiodistal length; W = buccolingual width; AW = anterior width; PW = posterior width.

Tooth	Mean (mm)	n	Observed range (mm)
P_1 L	25.2	5	22.3–28.5
P_2 W	15.7	5	13.8–19.1
P_3 L	22.45	4	20.0–24.5
P_3 AW	16.58	4	15.1–17.5
P_3 PW	18.23	4	16.0–21.4
P_4 L	22.9	6	20.2–25.2
P_4 AW	18.47	6	15.9–20.6
P_4 PW	18.7	6	15.90–22.0
M_1 L	24.7	7	20.00–26.9
M_1 AW	19.2	7	17.0–22.0
M_1 PW	20.4	7	17.5–22.8
M_2 L	25.87	7	22.5–29.0
M_2 AW	20.7	7	18.4–23.4
M_2 PW	20.3	7	17.8–22.8
M_3 L	27.3	7	25.4–32.2
M_3 AW	20.5	7	19.0–23.1
M_3 PW	18.6	7	17.0–21.7

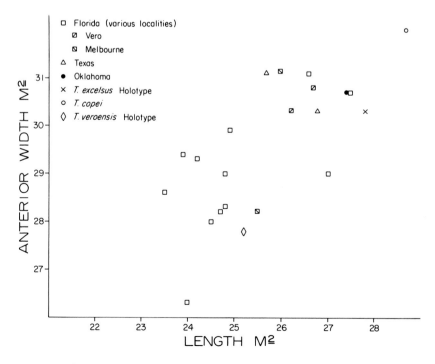

Fig. 7 Scatter diagram of length vs. anterior width of M^2 of various Pleistocene tapirs.

width of the M^2 (Fig. 7), however, shows the *veroensis-excelsus* sample to be divisible into two groups. An examination of the scatter diagram shows that the two groups do not coincide with the various samples but that some from each sample is represented in both groups. This seems to be best explained as reflecting sexual dimorphism.

Conclusions

(1) *T. haysii* and *T. tennesseae* are so inadequately based that they may be unidentifiable and, following Simpson (1945), are properly ignored.

(2) *T. copei* is so large that it does not overlap with the observed size-range of *T. veroensis* or any living species. Furthermore, its greater development of the protocone of P^1 increases the probability of the validity of the species.

(3) *T. californicus* and *T. merriami* are large and may be valid. However, more and better material is needed to validate these species.

(4) The strongest diagnostic feature of *T. excelsus* appeared to be the presence of a "sagittal table" rather than a medial sagittal crest. A series of Recent and fossil skulls, not available to Simpson in 1945, has demonstrated that the character of the sagittal table in the immature individual on which the species was based is a normal stage of growth for species with simple sagittal crests.

Therefore, nothing save the slightly large size and minor differences in the ratios of some teeth (which may also be related to the greater size) separates *T. excelsus* from typical *T. veroensis*. The dimensions of the dentitions of other fossils, including the Texas skull, Oesch's Missouri skull, Stovall and Johnston's Oklahoma skull, and a number of isolated Florida specimens, form a group intermediate in size between *T. veroensis* and *T. copei*. These are also intermediate in the development of the protocone of P^1. It seems probable that the size difference between *T. veroensis* (holotype and topotypes) and most other isolated finds of tapirs in the eastern half of the United States reflects no more than population and/or subspecific levels of separation. If larger samples from more localities demonstrate subspecific differences, the larger size may become known as *T. veroensis excelsus*.

Acknowledgments

We should like to express our sincere appreciation to the following for permission to examine material under their care: Charles Handley, United States National Museum; Horace G. Richards, Philadelphia Academy of Science; Karl Koopman, American Museum of Natural History; Thomas Patton and David Webb, Florida State Museum; Stanley Olsen, Florida Geological Survey; Thomas Pulley, Houston Museum of Natural Sciences; H.V. Andersen, Louisiana State University.

The work was initiated by mapping done for the United States National Park Service (Southwest Region) and carried out in part under National Science Foundation Grant GE-28598x.

Figs. 1 and 2 were prepared by Guinn Powell, Southern Methodist University. The preparation of Figs. 5–7 was made possible by the Owen-Coates Fund of the Geology Foundation, University of Texas at Austin.

Literature Cited

COPE, E.E.
    1899   Vertebrate remains from the Port Kennedy bone deposit. Jour. Acad. Nat. Sci.
            Philadelphia, 11: 193–267.

HAY, O.P.
    1920   Descriptions of some Pleistocene vertebrates found in the United States. Proc.
            U.S. Natl. Mus., 58: 83–146.

LEIDY, J.
    1860   Description of vertebrate fossils. *In* Holmes, F.S. Post-Pleistocene fossils of South
            Carolina, Charleston, 8, 9, 10: 99–122.

LUNDELIUS, E.L.
    1972   Fossil vertebrates from the late Pleistocene Ingleside Fauna, San Patricio County,
            Texas. Univ. Tex. Bur. Eco. Geo. Rept. Inv., 77: 1–74.

OESCH, R.D.
    1967   A preliminary investigation of Pleistocene vertebrate fauna from Crankshaft Pit,
            Jefferson County, Missouri. Natl. Speleogical Soc. Bull., 29(4): 163–185.

SELLARDS, E.H.
    1918   The skull of a Pleistocene tapir including description of a new species and a note
            on the associated fauna and flora. Florida State Geol. Surv. 10th and 11th Ann.
            Repts.: 57–70.

SIMPSON, G.G.
    1945   Notes on Pleistocene and Recent tapirs. Bull. Amer. Mus. Nat. Hist., 86(art. 2):
            37–81.

SIMPSON, G.G., A. ROE AND R.C. LEWONTIN
    1960   Quantitative Zoology, rev. ed. Harcourt, Brace, New York.

SLAUGHTER, B.H.
    1965   Preliminary report on the paleontology of the Livingston Reservoir Basin, Texas.
            Fondren Sci. Series, 10: 1–12.

STOVALL, J.W. AND C.S. JOHNSTON
    1934   *Tapirus haysii* of Oklahoma. Amer. Midland Nat., 15: 92–93.

Appendix

Table 4. Measurements of dentitions of Pleistocene tapirs from the United States.

ABBREVIATIONS

* = estimated measurement; FGS, Florida Geological Survey; USNM, US National Museum; UF, University of Florida; Tex. A & M, Texas A & M University; SMP-SMU, Shuler Museum of Paleontology, Southern Methodist University; HNSM, Houston Natural Science Museum; SM-UO, Stovall Museum, University of Oklahoma; CM, Central Missouri State College; AMNH, American Museum of Natural History; LSU, Louisiana State University; PAS, Philadelphia Academy of Science.

Table 4.

	Sante Fe. Fla. UF 2125	Arredondo UF 2560	Hornsby Spring UF 3270	Itchtucknee UF 11307	Itchtucknee UF 14064	New Apollo Bend UF 8225	Terra Cotta Bay UF 14064	Ingleside TMM 30967-1324	Ingleside TMM 30967-1237	Ingleside TMM 30967-193	Friesenhahn TMM 933-370	Missouri Oesch. 1967	Port Kennedy (T. copei) PAS 177
P_2 L		23.6	25.6		26.0		22.3	23.9	25.1	23.8		24.7	25.2
P_2 W		14.8	15.1		15.9		13.8	15.7	15.2	16.0		16.5	17.8
P_3 L		20.0	22.0		23.3			21.8	21.2	21.9		22.3	24.3
P_3 AW		15.1	16.6		17.1			16.5		16.1		16.8	26.9
P_3 PW		16.9	16.0		18.6			18.9	18.0	18.6		19.3	19.6
P_4 L	25.0	20.8	23.8		22.5		20.2	23.2			23.8	21.3	24.3
P_4 AW	17.5	17.5	19.0		20.3		15.9	19.8			16.3	20.4	19.6
P_4 PW	15.9	19.1	17.1		20.9		17.2	20.8			17.4	21.3	20.3
M_1 L		20.0	25.1		23.5	30.1	21.5	23.7		23.4	24.6	28.7	27.7
M_1 AW		17.5	20.9		19.7	21.8	18.2*	18.6		19.9	18.5	19.9	21.0
M_1 PW		17.0	18.3		19.2	22.0	17.4*	17.4	18.1		19.9	19.2	19.0
M_2 L	28.2	22.5			26.5		24.3	24.2	25.5	26.3	24.6	26.3	29.1
M_2 AW	21.6	18.4			21.4		18.9	20.3	21.4	20.7	20.4	21.8	21.9
M_2 PW	20.9	18.2			20.8		17.8	19.1	19.2	19.4	19.2	21.3	20.4
M_3 L		25.4		26.3			25.5	26.0					29.3
M_3 AW		19.2		19.8			19.8	17.6					21.5
M_3 PW		17.4		18.0			17.0						19.4

Table 4, continued.

	Brazos River Tex. A & M 2237	Forney SMP-SMU	Ingleside TMM 30967-1214	Ingleside TMM 30967-198	Ingleside TMM 30967-928	Livingston HNSM 160	Stoval, 1934	Missouri CM 159 Oesch. 1967	Simpson, 1945 (T. excelsus) AMNH 39406	Louisiana LSU	Port Kennedy Cave PAS 177	Port Kennedy Cave PAS 175
P^1 L	20.3		19.3			20.4	19.2	21.2			22.7	24.2
P^1 W	17.2		14.5			19.5	19.6	18.8			21.5	20.5
P^2 L	20.5					20.4	19.8	20.0			21.8	22.5*
P^2 AW	23.9					23.2	23.8	23.4			23.5	25.3
P^2 PW	24.2					25.9	24.2	25.0			25.3	26.6
P^3 L	21.5	23.5		21.5		20.6	21.2	20.8			22.5	23.0
P^3 AW	25.8	27.6		26.1		27.4	25.9	26.3			26.0	26.5
P^3 PW	25.0	27.0		25.8		27.6	25.7	26.3			25.9	26.5
P^4 L		24.5				21.0	21.9	20.0		24.2*	23.9	
P^4 AW		30.3				29.6	29.4	28.6		28.7	28.8	
P^4 PW		29.3				29.3	29.1	27.6		27.6	27.9	
M^1 L	22.6	25.0	23.1	25.6		23.4	23.9	22.7	25.0		27.1	
M^1 AW	25.2	28.9	26.3	27.7		28.3	29.2	27.3	26.4		28.9	
M^1 PW	24.0	27.1	24.0	25.7		25.6	27.2	25.0	25.1		26.1	
M^1 PW	24.0	27.1	24.0	25.7		25.6	27.2	25.0	25.1		26.1	
M^2 L	26.8				27.0	25.7	27.4	25.3	27.8	30.1	28.7	
M^2 AW	30.3					31.1	30.7	30.9	30.3	26.2	28.5	
M^2 PW	27.0				26.5	26.8	29.0	27.0	26.6	26.0	28.5	
M^3 L		28.5				23.7	27.3		26.5	26.5	28.0	
M^3 AW		32.9				27.6	29.5			31.5	31.1	
M^3 PW		29.1				22.3	29.1			25.5	26.6	

Table 4, concluded.

	Holotype *T. veroensis* FGS v277	Topotype FGS v4389	Topotype FGS 5447	Melbourne, Fla. USNM No. 3	Melbourne, Fla. USNM No. 2	Sante Fe UF 809	Arredondo UF 2560	Hornsby Spring UF 885	Hornsby Spring UF 886	Hornsby Spring UF 2889	Hornsby Spring UF 888	Itchtucknee UF 11310	Itchtucknee UF 14064	Pioneers Landing UF 11328	Branford IA UF 14056
P^1 L	17.5		19.0	19.2	19.2		18.2	18.0	18.5	19.5	19.5				
P^1 W	14.9		18.2	17.7	18.5		17.8	17.2	18.6	15.0	16.5				
P^2 L	18.7		20.5	20.8			19.0	19.2	21.1	19.4	20.6				
P^2 AW	20.9		21.3	23.2			21.3	21.2	23.2	19.6	20.9				
P^2 PW	22.8		24.0	24.7			24.1	22.8	25.8	23.4	24.1				
P^3 L	19.1		20.3				19.0	20.3	21.8	20.4	21.2			22.0	
P^3 AW	24.5		24.9				24.1	24.8	26.3	23.5	25.2			22.0	
P^3 PW	24.1		25.5				24.7	24.5	26.5	24.7	25.2			24.4	
P^4 L	20.7		20.9	22.5	22.3	22.0	20.1	21.0	22.3	22.0	21.5	21.0		21.9	
P^4 AW	26.9		27.6	28.4	26.3		26.9	26.8	28.6	24.2	27.3	27.8		25.0	
P^4 PW	26.2		27.5	27.0	24.1		25.9	26.3	28.1	22.5	27.3	26.3		25.7	
M^1 L	22.4	23.5	22.7		22.1	22.0	20.2	21.8	23.8	23.1	23.8	21.8	23.7	23.0	22.0
M^1 AW	26.8	28.4	26.4		26.5	25.7	25.5	25.2	27.7	25.9	26.9	26.0	27.1	26.1	24.2
M^1 PW	24.0	25.1	24.2		26.2		23.3	22.3	25.7	24.2	23.0	23.8	25.3	24.3	23.0
M^2 L	25.2	26.7	26.2	26.0	25.5	23.9	23.5	24.7	27.5	24.8	24.9	24.8	26.6	24.5	24.0
M^2 AW	27.8	30.8	30.3	31.1	28.2	29.4	28.6	28.3	30.7	28.3	29.9	29.0	31.1	28.0	26.3
M^2 PW	26.1	28.4	26.1	27.0	25.6	26.2	24.8	26.0	26.7	26.0	25.9	26.8	27.6	26.4	23.8
M^3 L	25.1	25.5		25.6	24.6		23.5	25.9				23.9	25.4	26.1	
M^3 AW	29.0	29.9		30.5	28.5		28.0		30.1			29.5	29.7	32.1	
M^3 PW	24.6	25.8		25.4	24.5		23.7	24.9	24.9			24.4	24.0	28.1	

The Oligocene Rodents *Ischyromys* and *Titanotheriomys* and the Content of the Family Ischyromyidae

Albert E. Wood
Professor of Biology, Emeritus, Amherst College,
Amherst, Massachusetts *and*
20 Hereford Avenue, Cape May Courthouse, New Jersey

Abstract

The Ischyromyidae (consisting of the genera *Ischyromys* and *Titanotheriomys*) and the Paramyidae have been considered distinct families. Black (1968) reduced *Titanotheriomys* to synonymy with *Ischyromys* and included the Paramyidae in the Ischyromyidae.

Features of the skull are discussed that show that there had been notable changes in the masseter muscle between the protrogomorphous paramyids and *Ischyromys*; even more pronounced ones characterize the sciuromorphous or myomorphous *Titanotheriomys*. Other changes in cranial muscles, nerves, and blood vessels reflected in the skull, together with differences of incisive histology, justify the recognition of *Titanotheriomys* as a valid genus and of the Paramyidae as a distinct family.

Ischyromys and *Titanotheriomys* were contemporaneous and sympatric during early and at least part of middle Oligocene time. If the ischyromyids were directly descended from paramyids, *Leptotomus* is considered the most probable known ancestor.

Introduction: Taxonomic History of Ischyromys *and* Titanotheriomys

The genus *Ischyromys* Leidy, 1856 is one of the better-known rodents of the North American Oligocene. Major recent studies of the genus have been those of Wood (1937), Howe (1966), and Black (1968). Wood described the skull and skeleton of *Ischyromys* in detail and recognized four species: *I. typus* Leidy 1856; *I. parvidens* Miller and Gidley, 1920; *I. pliacus* Troxell, 1922; and a new species, *I. troxelli*. He followed Miller and Gidley (1920) in recognizing *Titanotheriomys* (described originally by Matthew in 1910 as a subgenus of *Ischyromys*) as a valid genus, with the type species, *T. veterior*, from Pipestone Springs, Montana, and a new species, *T. wyomingensis*, from Beaver Divide, Wyoming. He recognized a number of characters separating the genus from *Ischyromys*, primarily those of the skull. Among these were not only a curved dorsal profile of the skull, considered by Black (1968, pp. 279–280) to result from crushing, but also differences in the masseter and temporal muscles and their attachment to the skull. Wood considered *Titanotheriomys* as primarily Chadronian, although he questionably indicated its presence in the Orellan (Wood, 1937, table, p. 262); *Ischyromys* he indicated as being present throughout the Oligocene.

Howe (1966) reviewed the middle and late Oligocene material of *Ischyromys* in the collections in the University of Nebraska. He recognized four successive species in Nebraska: *I. parvidens* from the lower Orella; *I. typus* from the middle Orella; *I. pliacus* from the upper Orella and perhaps basal Whitney; and *Ischyromys* sp. from the basal Whitney. He did not discuss *Titanotheriomys*.

Black (1968) concluded that the cranial differences between *Titanotheriomys* and *Ischyromys* that Wood had cited were all due to crushing, represented individual variation, or were non-existent; that there were no differences in the cheek-teeth except ones only detectable statistically; and that therefore *Titanotheriomys* was a synonym of *Ischyromys*. He reduced all previously described species to a "middle to possibly late Oligocene" species, *I. typus* (1968, p. 277), and an early to middle Oligocene species, *I. veterior* (including material previously referred to *I. parvidens*), and considered them derived from a common stock in the early Oligocene. He described a new species, *I. douglassi*, from the Chadronian of McCarty's Mountain, Montana, which was more primitive in its dentition than the other two. Black saw no significant differences in skull structure among *Ischyromys*, *Titanotheriomys* and paramyids, stating (1968, p. 275) that *Ischyromys* (including *Titanotheriomys*) "is only separated from late Eocene paramyines such as *Leptotomus*, *Thisbemys*, and *Rapamys* by a few details of cheek-tooth morphology," and (1968, p. 274) that "in the genus *Ischyromys* there is no indication of appreciable departure from the typical late Eocene paramyid morphology."

Therefore, he included the Paramyidae in the Ischyromyidae, the name that has priority. He seemed to have made the paramyids a subfamily of the Ischyromyidae (1968, p. 275), but his later classification (1971, pp. 182–183) makes it clear that he had intended to eliminate the Paramyidae, because he included the Paramyinae, Reithroparamyinae, and Prosciurinae as subfamilies of the Ischyromyidae, with the Manitshinae and Microparamyinae eliminated as separate taxa.

Since Black's two papers (1968, 1971) no detailed studies of these questions have been published. In his second paper, Black made a number of changes (based on the morphology of isolated cheek-teeth) in the allocation and synonymy of genera within the subfamilies of Paramyidae that had been recognized on the basis of cranial and mandibular characters (Wood, 1962). Table 1 contrasts Black's (1971) classification with that accepted in the present paper. Some authors have

Table 1. Comparison of the classifications by Black (1971) and Wood (current).

Black	Wood
Family Ischyromyidae	Family Ischyromyidae
Subfamily Ischyromyinae	*Ischyromys*
Ischyromys (incl. *Titanotheriomys*)	*Titanotheriomys*
	Family Paramyidae
Subfamily Paramyinae	Subfamily Paramyinae
Paramys, Thisbemys, Leptotomus (incl.	*Paramys, Thisbemys, Leptotomus,*
Tapomys), *Pseudotomus, Ischyrotomus,*	*Uriscus*
Manitsha, Mytonomys, Rapamys, Pseudo-	Subfamily Manitshinae
paramys, Plesiarctomys, Hulgana	*Pseudotomus, Ischyrotomus, Manitsha*
Subfamily Reithroparamyinae	Subfamily Reithroparamyinae
Reithroparamys (incl. *Uriscus*), *Franimys,*	*Reithroparamys, Franimys, Rapamys,*
Microparamys, Lophiparamys, Janimus	*Tapomys*
	Subfamily Microparamyinae
	Microparamys, Decticadapis, Loph-
	iparamys, Janimus
	Subfamily Ailuravinae
	Ailuravus (incl. *Maurimontia*), *Meldimys,*
	Mytonomys
	Subfamily Pseudoparamyinae
	Pseudoparamys, Plesiarctomys
Incertae sedis	*Incertae sedis*
Decticadapis, Ailuravus, Maurimontia,	*Hulgana*
Meldimys	
Subfamily Prosciurinae	Family Prosciuridae
Cedromus, Pelycomys, Plesispermophilus,	*Cedromus, Pelycomys, Plesispermophilus,*
Prosciurus, Spurimus	*Prosciurus, Spurimus*

followed Black in referring a variety of Eocene rodents to the Ischyromyidae; others have not. This situation can only lead to confusion.

Russell (1972, p. 28), in dealing with the rodents of Cypress Hills of Saskatchewan, described a new species *Ischyromys junctus*. In his discussion (pp. 29–30), he cited Black (1968) but then went on to say, "My own studies of specimens from Pipestone Springs in the Carnegie Museum and from the Vieja Group, preserved in the University of Texas, have led me to an opposite conclusion, i.e., that *Titanotheriomys* is a valid genus, distinguished by elongate molars, in which the cingular crests, especially the posteroloph, are widely separated from the main crests." In the same paper, Russell (pp. 19–20) continued to recognize the Paramyidae as a distinct family, although without discussion.

Wahlert (1974) investigated the cranial foramina of North American protrogomorphous rodents, and concluded that, on the basis of these features, the Ischyromyidae (*sensu* Wood, 1937) and Cylindrodontidae were quite similar to each other, and clearly distinct from the Paramyidae and Sciuravidae, which in turn were similar to each other. He made the Prosciurinae a distinct family, the Prosciuridae (Table 1).

My initial reaction had been to agree with Black's conclusions. However, a study of the rodents of the Vieja Group of Texas (Wood, 1974a), convinced me that I

had been correct in 1937 (pp. 194–195) that there were major differences in
the masseteric muscle between *Ischyromys* and *Titanotheriomys* that warranted
their generic separation, and that the differences between these (especially *Titanotheriomys*) and the Paramyidae were of sufficient importance to warrant the retention of the latter as a separate family. This opinion has been reinforced by
Wahlert's (1974) conclusions that there were important differences in the
cranial foramina of paramyids and ischyromyids.

I shall here present, first, the osteological and inferred myological differences
that I believe warrant the separation of *Ischyromys* and *Titanotheriomys* as distinct genera, and those that indicate that they are quite different from any known
members of the Paramyidae; second, other cranial and dental differences between
Ischyromys and *Titanotheriomys* on the one hand and the Paramyidae on the
other; third, a revised definition of the Family Ischyromyidae and its included
genera, *Ischyromys* and *Titanotheriomys*; fourth, a review of the geographic and
stratigraphic distribution of *Ischyromys* and *Titanotheriomys* and the resulting
implications for the phylogeny of the Ischyromyidae; and, last, a brief consideration of the probable ancestry of the Ischyromyidae. The differences between the
Paramyidae and Ischyromyidae are summarized in Table 2; those between
Ischyromys and *Titanotheriomys* in Table 3.

Abbreviations for collections mentioned in the text are: AEW, private collection
of A.E. Wood; AMNH, American Museum of Natural History, including the Frick
Collection; ANSP, Academy of Natural Sciences, Philadelphia; Basel, Natur-
historisches Museum, Basel; CM, Carnegie Museum; LACM (CIT), Los Angeles
County Museum (formerly the collections of California Institute of Technology);
TMM, Texas Memorial Museum of The University of Texas at Austin; UM, Montana State University at Missoula; and USNM, National Museum of Natural History
(formerly the United States National Museum).

Morphology of Ischyromys, Titanotheriomys *and the Paramyidae*

Musculature

MASSETERIC MUSCLE

The pattern of the masseteric muscle and the correlated modifications of the skull
have been considered of great importance in rodent classification at least since
Waterhouse (1839), and the use of this pattern was formalized as the basis of
rodent subordinal classification by Brandt (1855). Therefore, important taxonomic weight must be given to the marked difference in masseteric pattern between
the Ischyromyidae (as here defined) and the more primitive condition of the
Paramyidae.

In the Paramyidae, the ventral side of the zygomatic arch is horizontal. The scar
for the origin of the M. masseter lateralis is well shown in the lateral view (Fig.
1B; Wood, 1962, figs. 13B and 41C) running almost the length of the zygoma and
ending anteriorly level with the anterior cheek-teeth. The anterior limits of the

Table 2. Comparisons of the Paramyidae and Ischyromyidae.

Feature	Paramyidae	Ischyromyidae
Level of ventral margin of zyoma	Alveolar level of upper cheek-teeth	1/3 way from alveolar level to top of skull
Level of dorsal limit of M. masseter lateralis on zygoma	Slightly above alveoli of cheek teeth	Nearly 1/2 way up side of skull
Anterior face of zygoma	Forms angle of 75°–90° with forward projection of ventral surface of zygoma	Diagonal plate nearly continuous with ventral surface of zygoma; about 45° above alveolar plane
Origin of M. masseter superficialis	Rugosity forming anteromedian end of fossa for M. m. lateralis, lateral to P^4	Concave depression, distinctly separated from fossa for M. m. lateralis, in front of P^3
Muscle scars on snout lateral to diastemal ridges	Absent	Fossa for pars maxillaris anterior of M. buccinator
Premaxillary-maxillary suture on side of snout	Arches forward (except in *Manitsha*)	Vertical
Auditory bullae	Occasionally present; small except in *Reithroparamys* where they are inflated; no evidence of septa	Always present; large but not inflated; always septate
Orbital foramina	In primitive position	Displaced forward with respect to cheek teeth
Sphenofrontal foramen	On orbitosphenoid-alisphenoid suture	Often within alisphenoid
Foramen behind M^3 for descending palatine vein	Absent	Present
V^3 foramen between ectopterygoid flange and bulla	Absent	Present
Masticatory and buccinator foramina	Separate	Very close to, and usually fused with, V^3 foramen
Interorbital foramen	Absent	Present
Stapedial foramen	Present	Absent
Temporal foramen	Multiple and small	Single and large
Cheek teeth	Never fully crested; may have incipient mesoloph (-id)	P4/4–M3/3 always completely four-crested; no trace of mesoloph (-id)
Incisor enamel	Pauciserial where known	Always uniserial
Incisor enamel thickness	Thick to thin	Always thin
Cross-sectional shape of lower incisors	Round, oval, triangular with wide part in front, or egg-shaped with narrow part forward	Always egg-shaped with narrow end forward
Distribution of enamel on lower incisors	Usually limited to anterior face; in egg-shaped teeth, extends laterally past widest point	Extends laterally past widest point; covers about half the periphery of the tooth

Table 3. Comparisons of *Ischyromys* and *Titanotheriomys*.

Feature	*Ischyromys*	*Titanotheriomys*
Dorsal profile of skull	Always flat	Often arched
Anterior face of zygoma	Steep for an ischyromyid	Gently inclined
Origin of M. masseter lateralis	Ends anteriorly at strong crest curving medially on ventral side of zygoma, only slightly anterior to front of P^3	Extends lateral and forward of infraorbital foramen, at least as far as premaxillary-maxillary suture
Angle between central fibres of M. masseter lateralis and of M. masseter superficialis	About 60°	About 20°
M. masseter medialis penetrating infraorbital foramen	Never	Possible sometimes
Possible subdivision of infraorbital foramen	Never	Sometimes
Descriptive name of jaw musculature	Protrogomorphous	Sciuromorphous and/or myomorphous
Sagittal crest	Always well developed in adult	Usually absent; present in earliest known species (McCarty's Mt.), in one specimen from Pipestone Springs and two from Bates Hole
Temporal crests	Unite above eye in adult to form sagittal crest; in juvenile may not unite but are parallel	Usually (except McCarty's Mt.) curved with lyrate space between them; may unite near front of interparietal
Origin of pars maxillaris anterior of M. buccinator	A fossa, usually shallow, along lateral edge of palate well in front of P^3	A fossa, usually deep, sometimes closed ventrally, along lateral edge of palate in front of P^3
Bucco-naso-labialis	Scar for origin weak or absent	Scar for origin extends across entire premaxilla, almost continuous with that for masseter lateralis
Cranial foramina	no differences noted	
Incisors and cheek teeth	essentially no differences noted	

area of origin are always indicated by a strongly marked crest on the ventral side of the zygoma. This crest curves medially and ends at the base of the zygomatic arch (Fig. 1c). The anterior part of the fossa of the M. masseter lateralis is only slightly above the plane through the alveoli of the upper cheek-teeth, although the posterior parts of the fossa are somewhat higher (Fig. 1b). There is usually a clearly developed rugosity at the anteromedian end of the fossa, marking the origin of the tendon of the M. masseter superficialis (Fig. 1c; Table 2).

In all paramyids in which this area is preserved (except *Manitsha tanka*, which is, in respect to the masseteric fossa, *sui generis*), there is a large angle between the anterior part of the fossa for the M. masseter lateralis and the slope of the zygoma below the infraorbital foramen. Measurements of the angle taken from illustrations given by Wood (1962) range from 75 to 90 degrees above the horizontal in all paramyids (Table 2).

In *Leptotomus grandis*, although the angle between the two surfaces is over 75 degrees, the space between the infraorbital foramen and the angle at the front of the masseteric fossa is unusually large, owing in part to the retreat of the anterior margin of the fossa for the M. masseter lateralis, which lies above the middle rather than the front of P^4 (Wood, 1962, fig. 26). This increased space between the muscle scar and the infraorbital foramen led Wood (1962, p. 82) to suggest "the possibility that this space between the foramen and the zygoma was in the process of expanding, which could well have been a prerequisite for the 'sciuromorph' type of zygomasseteric structure." Although this statement concerned only the *space* between the infraorbital foramen and the area of origin of the masseter, it apparently was not clear, as it confused Black (1968, p. 275), who erroneously stated that "The migration of the masseter onto the forward face of the zygoma ... does not appear to be greatly different from the conditions discussed by Wood ... for *Leptotomus*, particularly *Leptotomus grandis*." (Black very kindly allowed me to read the manuscript of his 1968 paper in late 1967. I sent him a number of comments [pers. comm. lit., December 7, 1967] in which I pointed out instances where he was mistaken or where I thought he had not interpreted my previous remarks correctly, including this one. These comments apparently reached him too late for inclusion in his paper. Likewise, he has read an early draft of this paper, and disagrees with most of its interpretations and conclusions.)

In other species of *Leptotomus*, e.g. *L. bridgerensis* (Wood, 1962, figs. 31b; 33a), the masseteric fossa has not shifted backward, and ends near the front of P^4 or even farther forward (Wood, 1962, fig. 35b); the space between the infraorbital foramen and the muscle scar is correspondingly smaller. *Thisbemys* shows some arching of the ventral surface of the zygoma, raising the lateral part of the masseteric fossa above the level of the cheek-teeth (Wood, 1962, figs. 36a; 37j), but in a manner more like that in *Manitsha* than in *Ischyromys*. In the earliest known rodent skull, *Franimys*, the fossa for the origin of the M. masseter lateralis reaches appreciably forward of P^3 (Wood, 1962, fig. 48b), and in *Ischyrotomus* it reaches to the front of P^3 (Wood, 1962, figs. 68, 71). The masseteric fossa of *Manitsha* (Wood, 1962, figs. 80–81) is very distinctive, with a vertical anterior end and with the zygoma apparently raised well above the cheek-tooth level. The anterior, vertical part of the fossa is in front of P^3, resulting in a very considerable anteroposterior component for the M. masseter lateralis.

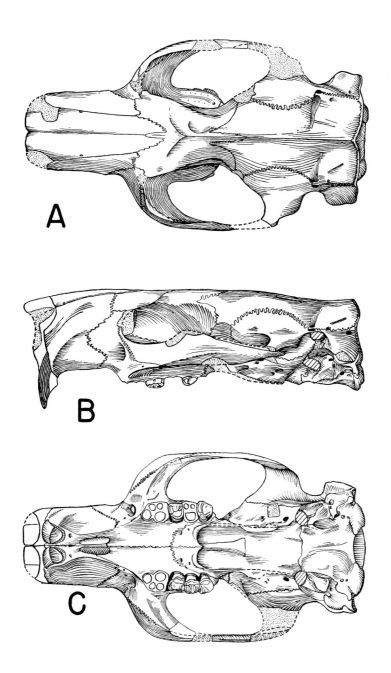

Fig. 1 *Paramys delicatus*, AMNH 12506, skull, × 1. (From Wood, 1962, fig. 2).
    A. Dorsal view.
    B. Lateral view.
    C. Ventral view.

In *Ischyromys*, by contrast, the zygomatic arch, although horizontal, has been elevated considerably above the alveolar border of the cheek-teeth. Thus the strong crest that marks the dorsal limit of the origin of the M. masseter lateralis and begins well back on the zygoma is closer to the mid-height of the skull than to its ventral margin (Fig. 2B, Table 2). The area of origin of the anterior end of the M. masseter lateralis on the ventral surface of the zygomatic root neither is horizontal in lateral view, as in paramyids, nor does it form a large angle with the slope of the zygomatic plate between the masseteric fossa and the infraorbital foramen (Fig. 2B; Wood, 1940, pl. 34, fig. 1). Rather there is a continuous diagonal zygomatic plate, higher at the anterior end, elevated about 45 degrees above a plane through the alveolar border of the upper cheek-teeth, and sloping almost uninterruptedly from the infraorbital foramen to the posterior margin of the maxillary root of the zygoma (Fig. 2B).

The situation in *Ischyromys* is very different from that observed in any paramyid. Wilson (1940, p. 94) also noted this difference, and commented that *Griphomys* "exhibits a tilted zygomatic plate similar to that possessed by most modern rodents and unlike the corresponding structure in the primitive protrogomorph members of the order. In view of its fragmentary nature, possibly the plate is an extreme example of the type possessed by the protrogomorph rodent *Ischyromys* ..." *Ischyromys* always has a prominent anterior margin to this fossa, curving medially from the crest along the lateral side of the zygomatic arch, and clearly separating the fossa for the M. masseter lateralis from the slope below the infraorbital foramen (Fig. 2C, Table 3). In at least some specimens, a faint continuation of this crest extends backward just lateral to the anterior end of the cheek-teeth. The transversely directed anterior end of the crest lies about halfway between the posteroventral margin of the infraorbital foramen and the posterior margin of the zygomatic root, lateral to the premolars. Unquestionably this crest was the origin of the M. masseter lateralis.

There is no knob adjacent to the fossa for the M. masseter lateralis for the tendon of the M. masseter superficialis, as in the Paramyidae; rather, anterolateral to the alveolus of P^3 (Table 2), there is a depression, which is quite variable in size and character, but which is often fairly deep and rugose (Fig. 2C; Burt and Wood, 1960, fig. 1B). Probably this marked the attachment of the tendon of origin of the M. masseter superficialis (Black, 1968, p. 289, stated that this depression was "the place of origin of part of the masseter"). If this is correct, the origin of the M. masseter superficialis must have migrated, in *Ischyromys*, a considerable distance anteriorly from its position in the Paramyidae next to the origin of the M. masseter lateralis.

These changes from the Paramyidae to *Ischyromys* suggest to me that other important changes had taken place in the structure and arrangement of the parts of the masseteric muscle and, therefore, in the gnawing mechanism (Fig. 4).

A more advanced stage of development of the masseteric muscle is shown in the skulls of *Titanotheriomys*. In these, the arrangement of the masseteric scars is similar *in some respects* to that in *Ischyromys*, but is clearly much advanced in a sciuromorphous (or even myomorphous—see below) direction. One important item for the classification of these rodents is that the arrangement in *Titanotheriomys* could readily have been derived from that in *Ischyromys*, but that the reverse

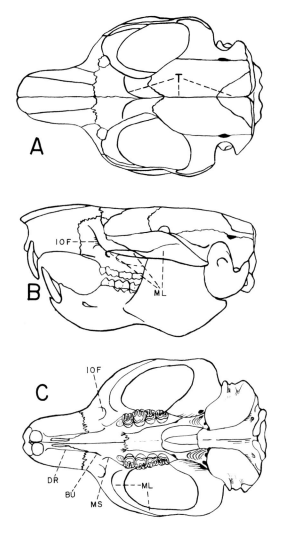

Fig. 2 *Ischyromys typus*, composite restoration, × 1.
    A. Dorsal view of skull.
    B. Lateral view of skull and mandible.
    C. Ventral view of skull.
    Abbreviations: BU — fossa for pars maxillaris anterior of M. buccinator; DR — diastemal ridge; IOF — infraorbital foramen; ML — fossa for origin of M. masseter lateralis; MS — fossa for origin of M. masseter superficialis; T — limits of M. temporalis.

would have been very unlikely. As in *Ischyromys*, the ventral surface of the zygoma is well above the level of the cheek-teeth; the anterior face of the zygomatic arch, including the infraorbital foramen and the plate ventral and posterior to the foramen, is inclined to the horizontal at about the same 45-degree slope as in *Ischyromys* (Fig. 3B, Table 3).

However, the scar curving medially across the plate ventral and posterior to the foramen in *Ischyromys* and marking the anterior end of the M. masseter lateralis is absent in *Titanotheriomys* (cf. Figs 2c and 3c). Instead, the dorsal (or lateral) boundary of the M. masseter lateralis developed along the posterior portions

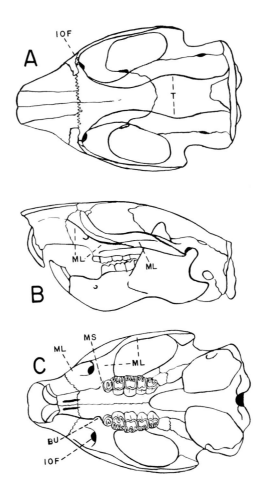

Fig. 3 *Titanotheriomys wyomingensis*, AMNH 14759, skull and mandible, × 1.
    A. Dorsal view.
    B. Lateral view.
    C. Ventral view.
    Abbreviations as in Fig. 2 (Redrawn from Wood, 1937, pl. 27).

of the zygoma in a manner similar to that in *Ischyromys*, and continued forward, lateral to and above the the infraorbital foramen (Figs. 3B, C–ML; 6A, B–ML; Table 3). This development was originally pointed out by Matthew (1910, p. 63), who stated (in his definition of the subgenus *Titanotheriomys*) that the "superior border of origin of the masseter extended forward on muzzle ..." In front of the foramen, the boundary is not so sharp as farther back, indicating a gradual attenuation of the anterior portions of the M. masseter lateralis. Nevertheless the scar extends upward across the dorsal root of the zygoma and forward at least to the premaxillary-maxillary suture near the point where this suture turns ventrally down

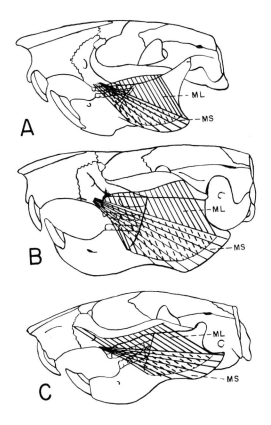

Fig. 4 Restorations of Mm. masseteres laterales and Mm. masseteres superficiales of selected rodents.
    A. *Ischyrotomus.*
    B. *Ischyromys.*
    C. *Titanotheriomys.*
    Abbreviations: ML — M. masseter lateralis; MS — M. masseter superficialis.

the side of the snout, well forward of the infraorbital foramen (Fig. 6B; Table 3). Black (1968, p. 282) was, I believe, entirely mistaken when he said that "in no specimens available does the masseter extend above the infraorbital foramen. This groove is absent in two of the Pipestone skulls ..."; or when he stated (1968, p. 289) "no specimens show any indication that the masseter has spread off the ventral border of the zygoma."

Such a forward displacement of the origin of the M. masseter lateralis seems clear to me, even in those skulls of *Titanotheriomys*, where Black specifically denied its presence. Unfortunately, I have been unable to convince Black of the presence of the fossa for the M. masseter lateralis, lateral and anterior to the infraorbital foramen, even when we were looking at the same specimen and I was tracing the outline of the fossa with a pointer. I have, however, convinced a number of other observers of my accuracy. Even when Black and I agree on the structure, we generally disagree on its myological significance.

This anterior position of the area of origin of the M. masseter lateralis seems to me to be observable on all skulls that I refer to *Titanotheriomys*: i.e. all those from Pipestone Springs and McCarty's Mountain, Montana; 14 skulls found between the 250-foot level and the 305-foot level of Bates Hole, Wyoming, in the Frick Collections, American Museum; AMNH 14579 and 14966 from Beaver Divide, Wyoming; and TMM 40289–91 from the Ash Spring local fauna of the Vieja Group of Texas.

The only logical interpretation of this condition in *Titanotheriomys* is that the origin of the M. masseter lateralis has spread forward off the zygoma, in an apparently sciuromorphous manner (Wood, 1937, pp. 194–195). Such a migration of the origins of the muscles, although observable in all skulls or appropriate fragments of *Titanotheriomys* that I have seen, did not always produce equally marked ridges on the snout in front of the infraorbital foramen, which may be why Black sometimes doubts the presence of such ridges.

In *Titanotheriomys*, there is always a pit in front of P^3, similar to that in *Ischyromys*, and likewise presumably marking the attachment of the tendon of the M. masseter superficialis, as there is no other area where this muscle could have arisen (Table 3 and Fig. 6A, B; MS).

Black believes that the M. masseter lateralis varied within populations from what I consider the *Ischyromys*-type to what I consider the *Titanotheriomys*-type. In my opinion there is no reported instance of a Chadronian collection, from restricted temporal and geographic limits, that shows a transitional series of specimens. Therefore, since the structure of the M. masseter lateralis does not vary significantly within a single population, and since it seems highly unlikely that the forward movement of the origin of the M. masseter lateralis off the zygoma to a position lateral to and above the infraorbital foramen should be a secondary sexual character, I believe it must be of taxonomic value.

BUCCINATOR MUSCLE

In all paramyid skulls with which I am familiar, except that of *Franimys* (Wood, 1962, fig. 48B), there are prominent ridges that run along what was undoubtedly the alveolar margin of the pre-rodent palate, from the alveoli of P^3 to or nearly to the alveoli of the incisors, and separate the palatal and lateral surfaces of the snout (Fig. 1c and Wood, 1962, figs. 13c, 31c, 37k, 41b, 68b, 71b and 81). These are referred to below as "diastemal ridges" (Figs. 2c, DR; 5A and B, DR; 6A, DR; and 7, DR). In the Paramyidae, I have not observed fossae lateral to these ridges for the attachment of muscles on the sides of the snout (Table 2).

In *Ischyromys*, there is the same diastemal ridge marking the lateral border of the palate (Figs. 2c, DR and 5B, DR), although it is often much fainter than in paramyids. However, a distinct fossa of variable size (not noted in any paramyid) lies anterior to the pit for the attachment of the M. masseter super-ficialis and just buccal to the posterior end of the diastemal ridge (Figs. 2c, BU

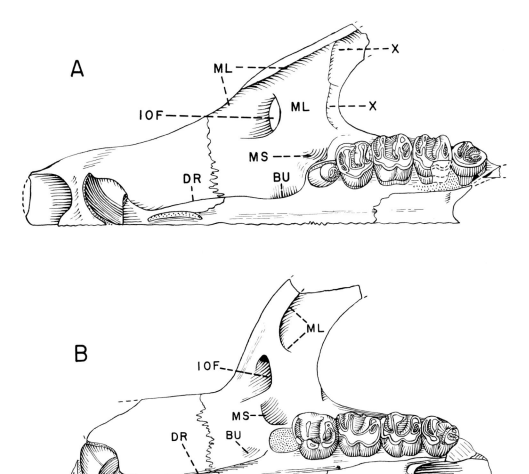

Fig. 5 Left halves of palate and snout of ischyromyids from McCarty's Mountain, Montana, and Slim Buttes, South Dakota.
   A. *Titanotheriomys douglassi*, CM 1053, partly restored from the opposite side.
   B. *Ischyromys typus*, CM 9764.
   Abbreviations as in Fig. 2; x — possible subdivisions of the fossa for the M. masseter lateralis or possibly breaks in the bone.

and 5B, BU; Table 2). This seems to mark the area of origin for a facial muscle, probably the pars maxillaris anterior of M. buccinator (D.J. Klingener, pers. comm.; see Meinertz, 1943, p. 372, fig. 20, for a similar condition in *Marmota*, in which the masseter clearly has nothing to do with this area).

In *Titanotheriomys*, there is a fossa, similar to but larger than that in *Ischyromys* described above, medial and anterior to the fossa for the M. masseter superficialis and just lateral to the posterior end of the diastemal ridge (Fig. 3C, BU). This fossa varies within the genus (and apparently, within populations), sometimes being large or extremely large and deeply excavated into the side of the snout, facing, in essence, laterally (Table 3). In the latter case, the diastemal ridge forms a prominent shelf ventral to this fossa (Fig. 6A, B, BU).

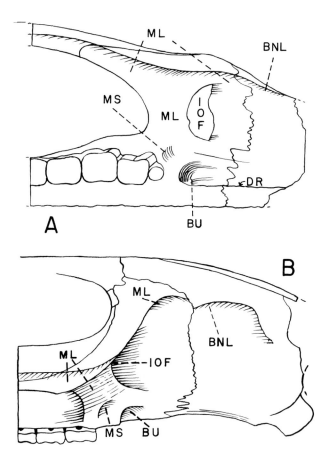

Fig. 6A–B *Titanotheriomys veterior.*

    A. Oblique view of the right side of the palate and zygomatic arch, AMNH F:AM 79316, × 3. From 15 feet (= 5 m) above 285-foot level, Bates Hole, Wyoming.

    B. Lateral view of right side of snout, AMNH 42993, × 3. From 295-foot level, Bates Hole, Wyoming.

    Abbreviations as in Fig. 2; BNL — fossa for M. bucco-naso-labialis.

BUCCO-NASO-LABIALIS MUSCLE

I have not detected scars for the attachment of any facial muscles on the dorsal portions of the lateral surfaces of snouts that I refer to *Ischyromys* or the paramyids. However, in *Titanotheriomys* there is a curving ridge, marking another muscle attachment, that extends forward across the premaxillary and crosses the elevation that marks the course of the incisive alveolus (Table 3). In ventral view this ridge seems to be continuous with that on the maxillary marking the dorsal edge of the M. masseter lateralis (Fig. 6A, BNL), but in lateral view it is clear that there is an abrupt discontinuity between these ridges (Fig. 6B, BNL). The premaxillary ridge seems certainly to be the attachment of a muscle other than the M. masseter lateralis. Probably it was the origin of the M. bucco-naso-labialis (D.J. Klingener, pers. comm., Sept. 19, 1969).

Some specimens, however, show a remarkable variant. If the skull of AMNH F:AM 79312 is held at the correct angle to reflect the light from the lateral surface of the snout, a faint ridge can be seen as a posteroventral continuation of the scar interpreted as the origin of the M. bucco-naso-labialis. This posteroventral extension reaches the dorsomedial corner of the infraorbital foramen (Fig. 7; X) and suggests that the structure that reached the premaxillary ridge had passed through the infraorbital foramen. Dr. Richard H. Tedford independently observed this and pointed it out to me. This raises the very interesting possibility that part of the surface of the snout had been invaded by a slender portion of the M. masseter medialis which passed through the infraorbital foramen (Table 3). I feel no confidence in this interpretation, although the morphological facts on AMNH F:AM 79312 are unquestionable as indicated.

If the scar on the premaxilla represents the origin of the M. masseter medialis, there might be some indication of a separation between that part of the infraorbital fenestra that transmitted the muscle and that part reserved for the nerve and blood vessels. I have seen only one specimen (AMNH F:AM 79316, Fig. 6A) that shows any trace of such a separation; it has a small bony protuberance on the ventral margin of the infraorbital foramen (Table 3). In my opinion the infraorbital foramen averages slightly larger in *Titanotheriomys* than in *Ischyromys*, which would imply that, if the foramen in the former had been invaded by the M. masseter medialis, only a very thin slip was involved.

TEMPORALIS MUSCLE

Typically, in *Ischyromys*, the M. temporalis occupied a broad stretch of the skull, its anterior fibres arising from the curved supraorbital ridge (Fig. 2A, T); this reaches the margin of the orbit at a faint rugosity that probably marks the approximate posterior end of the eye. The two crests curve abruptly mediad to unite in a sagittal crest at the level of the postorbital constriction. The sagittal crest then continues to the posterior end of the skull. This pattern agrees with that in the skulls of most paramyids (Fig. 1A; Tables 2–3).

In *Manitsha, Ischyromys horribilis* and *I. petersoni* (but not *I. oweni*), and in *Leptotomus costilloi* and *L. bridgerensis*, the supraorbital crests unite just behind the postorbital constriction. The supraorbital and temporal crests of *Reithroparamys* (Wood, 1962, fig. 41A) are separate for their entire length, converging gradually until they are very close together just in front of the anterior tip of the

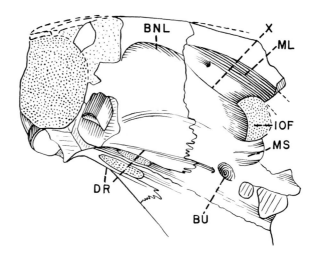

Fig. 7 Snout of *Titanotheriomys veterior*, AMNH F:AM 79312, Bates Hole, Wyoming. Oblique
antero-latero-ventral view.
Abbreviations as in Fig. 2; BNL — fossa for M. bucco-naso-labialis; x — ridge possibly
marking limits of M. masseter medialis.

interparietal. These variations of the M. temporalis in paramyids are not associated
with any observed differences in the masseter. Thus, in all paramyids and in
Ischyromys, the M. temporalis was very large, and there were many anterior fibres
that pulled the jaw forward, as well as more posterior fibres that pulled the jaw
backward and elevated it, pivoting it around the condyle.

The only skull with an *Ischyromys*-type of masseteric scar that I have seen that
does *not* have a sagittal crest is a juvenile skull (TMM 41216-10), still retaining
dP$^{3-4}$, from the Porvenir local fauna of the Vieja Group of Texas (Table 3). In
this specimen, the supraorbital crests curve smoothly medially, as in other skulls
of *Ischyromys*, but they do not quite meet to form a sagittal crest in the preserved
part of the skull, which ends approximately at the rear of the frontal (Wood,
1974b, fig. 8A). Although the crests are separate, they are completely parallel and
only about a millimetre apart. In this instance, I believe that the small size of the
M. temporalis was merely a juvenile character, and that adults of this population
would have had the sagittal crest normal in *Ischyromys*. This is probably not a
primitive character (even though this is probably the earliest ischyromyid skull
known), since sagittal crests are present in paramyids the size of *Ischyromys*.

The sagittal crest of *Titanotheriomys* may or may not be like that of *Ischyromys*.
In *T. veterior*, the supraorbital crests arise at the edge of the orbit, in the same
place as in *Ischyromys*, and likewise curve medially but rarely meet (Fig. 3A, T).
Posteriorly the crests diverge over the middle of the parietal before reconverging
farther posteriorly, but usually without meeting. As a result, there is a very prominent lyrate area on the dorsum of the skull that had no muscular covering and
lay between the fossae for the two temporal muscles (Fig. 3A). Both the anterior
and the posterior portions of the M. temporalis must have been smaller than in
Ischyromys.

One skull (CM 9058) of *T. veterior* from Pipestone Springs, Montana, and two (AMNH F:AM 79313 and F:AM 79318) from Bates Hole, Wyoming, have the typical forward displacement of the masseter that characterizes *Titanotheriomys*, but the temporal crests unite to form a sagittal crest as in *Ischyromys* (Black, 1968, p. 282). Sagittal crests are likewise present on all five Carnegie Museum skulls or skull fragments of the ischyromyid from McCarty's Mountain, considered below to be *Titanotheriomys douglassi* (Black).

The conformation of the sagittal crest, along with other characters, was used by me to separate *Titanotheriomys* from *Ischyromys* (Wood, 1937, p. 193). Although it is obviously not an absolutely diagnostic character at the generic level, I know of no ischyromyids with lyre-shaped temporal crests that did not also have the *Titanotheriomys*-type of masseter; every adult skull that I have seen with an *Ischyromys*-type of M. masseter lateralis has a sagittal crest.

Black (1968) argued that the differences in sagittal or temporal crests that I consider indicative of *Ischyromys* or of *Titanotheriomys* are merely sexual or age variations, reflecting differences in the size of the M. temporalis, which "is known to vary considerably within populations" (p. 282). One skull (CM 9058) from Pipestone Springs that possessed the *Titanotheriomys*-type of masseter had a sagittal crest, whereas the others did not; thus Black was clearly correct that there is variation in this feature. However, the presence of sagittal crests in all the specimens from McCarty's Mountain and their absence in a score of other *Titanotheriomys* skulls, except the three cited above, suggests that the crests were primitively present in the genus and progressively lost. The very distinct condition in the juvenile skull from the Vieja shows that the ontogeny of *Ischyromys* did not pass through a stage resembling the adult *T. veterior*.

In groups with strong sexual dimorphism (e.g., Pongidae) variation in the sagittal crest is often a secondary sexual character. The presence or absence of a sagittal crest, however, seems to be a valid criterion for separating some genera of foxes (e.g., *Urocyon* and *Vulpes*), and rodents notoriously show minimal sexual dimorphism. Since (with the exception of the skulls from McCarty's Mountain and the three variant skulls cited above) the differences in the M. temporalis are correlated with differences in the development of the M. masseter, I consider them significant.

Cranial Foramina

A recent investigation by Wahlert (1974) indicates that the cranial foramina of *Ischyromys* and *Titanotheriomys* are essentially identical, are quite different from those of paramyids and sciuravids, but, rather surprisingly, are quite similar to those of cylindrodonts. The following discussion is paraphrased from his results.

The snout of ischyromyids is elongate and forms about a quarter of the skull length compared with a fifth in paramyids. This has been accompanied by a backward displacement of the cheek-teeth from their positions in the Paramyidae. The orbital foramina seem to have retained their original positions, and so are apparently displaced forward with respect to the cheek-teeth (Table 2). The foramina involved include the nasolacrimal, the sphenopalatine (through which the descending palatine artery and nerve enter the skull), the posterior palatine (through which they leave it), the optic, the ethmoid, and the sphenofrontal. In the

Ischyromyidae, this last foramen is often contained entirely within the orbito-sphenoid, whereas, in paramyids, it lies on the orbitosphenoid-alisphenoid suture.

In the Ischyromyidae, certain bony processes have grown to adjacent areas of bone, completing foramina not present as such in the Paramyidae. The posterior process of the maxillary behind M^3 has grown backward to unite with the palatine, closing the posterior maxillary notch of paramyids to form a foramen that prob-ably transmitted the descending palatine vein. The large external pterygoid flange in ischyromyids grew back to unite with the large bulla, surrounding the course of the cranial nerve[3], and forming what Wahlert terms the V^3 foramen (Table 2).

The masticatory and buccinator foramina, separate in paramyids, are usually fused with the V^3 foramen in ischyromyids (Table 2). If they are present as separate foramina, they are very close to the V^3 foramen and the channels leading to them are very short. Foramina are present that do not occur in paramyids, including the interorbital foramen, the post-alar fissure and the squamoso-mastoid foramen.

There is no stapedial foramen in ischyromyids, presumably indicating the loss of the stapedial artery, so that the ophthalmic artery would have arisen from the internal maxillary (Table 2). The postglenoid foramen is similar in length to that in paramyids, but is much narrower, and the major venous flow had been shifted away from it. The temporal foramen is single and large, rather than multiple and small as in paramyids. The double hypoglossal foramen, opening into a pit, is unlike anything among the paramyids.

Ischyromyids apparently lack the orbital portion of the palatine. There is a depression for the rectus muscles in the orbitosphenoid in front of the optic foramen; the attachments of these muscles have not been identified in paramyids.

Cranial Sutures

There is a forward arching of the premaxillary-maxillary suture on the side of the snout in all paramyids where this region is known (Fig. 1B), except *Manitsha*, whereas the suture is nearly vertical in the ischyromyids (Figs. 2B, 3B, 4, 5, 6. and 7). I am ignorant of the significance of this difference. The condition in the paramyids is undoubtedly the more primitive.

In most paramyids, the nasals are long, extending well back of the front of the orbit, whereas the premaxillaries generally end in front of the orbit (Fig. 1A; Wood, 1962, figs. 31A, 35B, 36A, 65A, 68A and 71A). In *Franimys* (Wood, 1962, fig. 48A) both bones extend well behind the front of the orbit. The nasals extend behind the premaxillae but end near the front of the orbit in *Leptotomus costilloi* (Wood, 1962, fig. 24A), and only in *Reithroparamys* (Wood, 1962, fig. 41A) and *Plesiarctomys* (Wood, 1970, fig. 1B) are the rear ends of the premaxillae and nasals essentially on a common transverse line. In most Ischyromyidae (Figs. 2A, 3A) the posterior ends of the premaxillae and nasals form a nearly straight line, slightly anterior to the front of the orbit, although occasionally the nasals extend slightly farther posteriorly, but apparently never so far as in the paramyids.

The anterior end of the palatine crosses the palate "on a line between P^4 and M^1" (Black, 1968, p. 289) or even farther forward (Fig. 2C, 5B) in the ischyro-myids. This is a slight, but probably not very significant, difference from paramyids, in which the suture seems to reach the front of M¹ only in *Plesiarctomys* (Wood, 1970, figs. 2A, 7B), and is often considerably farther to the rear.

Auditory Bulla

Black stated (1968, p. 289) that "The foramen ovale opens just anterior to the bulla in *Ischyromys*, as it does in *Paramys*", an obvious *lapsus* as the bulla is known in no specimen of *Paramys*. In ischyromyids, the long axes of the bullae converge at a point near the posterior end of the bony palate. Among paramyids, the bulla of *Leptotomus sciuroides* is "rather similar to that of *Ischyromys*, but smaller [proportionately] than that of *Titanotheriomys*" (Wood, 1962, p. 99), but the area including the foramen ovale is not preserved.

In *L. sciuroides*, the point of convergence of the bullar axes is about half-way between the front of the bullae and the rear of the palate. A large, elongate bulla was present in the subhystricognathous *Reithroparamys*, also somewhat like that of ischyromyids, but again the foramen ovale cannot be seen. The convergence of the bullar axes is as in *L. sciuroides*. The only other paramyid in which the bulla is preserved is *Ischyrotomus oweni*, in which one specimen, USNM 17160, has a small, non-co-ossified bulla on the right side, displaced forward and rotated from its normal position. The foramen ovale is in front of, but rather far from, the bulla, and conditions are not very much like those in ischyromyids. The bulla of the ischyromyids is highly septate, with a large number of septa extending about a third of the way into the bullar cavity from all sides, and leaving only the central third of the cavity undivided (cf. CM 17453, *T. veterior*, Pipestone Springs).

As far as I am aware, there is no trace of septa in the bulla of any paramyid (Table 2). Because so few paramyid bullae are known, I feel that little weight can now be given to these structures in determining paramyid-ischyromyid affinities. However, they certainly do not support the close affinity assumed by Black.

Shape and Histology of Incisors

In all ischyromyids, the enamel of both upper and lower incisors is thin. The lower incisors have a very narrow anterior face, with an ovate cross-section that gradually widens to a level well toward the rear. The enamel extends, on the lateral side, to or behind the widest part of the tooth (Table 2). This is best seen in cross-sections (Fig. 8D, E) rather than in lateral or occlusal views. The upper incisor does not have so distinctive a cross-sectional shape, and the enamel does not extend so far around the lateral side of the tooth.

The lower incisors of most paramyids are quite different from those of ischyromyids (Fig. 8A). The greatest similarities to the ischyromyids are the conditions in various species of *Leptotomus* (Fig. 8C; Wood, 1962, figs. 22H, I, 23J, 30B–E, 33E, 34F), although at least some specimens referred to *Rapamys* have rather ischyromyid-like incisors (Fig. 8B) associated with complex upper molars (Black, 1971, p. 191). The striking similarity of the lower incisors is one reason that Wood (1962, p. 72) suspected a special relationship between *Leptotomus* on the one hand, and *Ischyromys* and *Titanotheriomys* on the other.

The enamel histology of the incisors of rodents is a most important taxonomic character, but one which is little understood functionally. In both *Ischyromys* and

Titanotheriomys the enamel is uniserial (Korvenkontio, 1934, p. 118) as in most post-Eocene non-hystricognathous rodents (Wahlert, 1968, pp. 15–17), whereas all paramyids (cited by Wahlert, 1968, p. 14, as Ischyromyidae, following Black, 1968) are pauciserial (Table 2). Uniserial enamel is present in *Cylindrodon* and *Prosciurus* (Wahlert, 1968, p. 15), which supports Wahlert's (1974) conclusion, based on cranial foramina, that the Ischyromyidae and Cylindrodontidae are specially related, and that *Prosciurus* and its relatives deserve familial recognition and are closer to the Sciuridae than to the Paramyidae. Although all paramyid incisors that have been sectioned show the primitive pauciserial condition, unfortunately no specimens of *Leptotomus* and only two post-Bridgeran paramyids (*Ischyromys petersoni*, Uintan, and *Manitsha* sp., Vieja Chadronian) have been studied (Wahlert, 1968, p. 14).

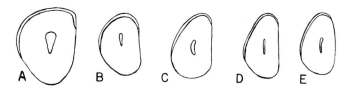

Fig. 8   Cross-sections of rodent left lower incisors, × 5.

  A. *Paramys delicatior*, ANSP 10275, holotype; a typical paramyid (From Wood, 1962, fig. 10c).
  B. *Rapamys fricki*, LACM (CIT) 2181, holotype (From Wood, 1962, fig. 52E).
  C. *Leptotomus parvus*, AMNH 12519, holotype (From Wood, 1962, fig. 30B).
  D. *Titanotheriomys veterior*, AEW 9676, Pipestone Springs, Montana.
  E. *Ischyromys* cf. *typus*, AMNH 42988, reversed, Chadronian, Douglas, Wyoming.

Lack of Cheek-Tooth Differences

Black (1968, pp. 290–300) indicated that there are no detectable cheek-tooth differences among middle Oligocene ischyromyids and lower Oligocene ones except for those from McCarty's Mountain, so he reduced the Oligocene ischyromyids to three species: *I. douglassi* from McCarty's Mountain; *I. veterior* from other Chadronian localities and from some Orellan ones; and *I. typus* from the Orellan (and perhaps the Whitneyan). Moreover, Black (1968, pp. 292–297) stated that the differences between the last two species are merely statistically detectable differences of size. Russell (1972, pp. 29–30), as pointed out above, disagreed with this, and considered that the cheek-teeth of *Titanotheriomys* were more elongate and had more divergent cingular crests than did those of *Ischyromys*. I have not reviewed all the abundant material of Oligocene ischyromyids in sufficient detail to permit me to have an intelligent opinion as to whether Black or Russell is correct.

Concerning his conclusions as to the synonymy of various described species, Black (1968, p. 278) was correct that there is no interparietal on the holotype of *I. troxelli*; it was, however, present in 1936 when I wrote my account of that form. He was mistaken in his statement that the interparietal was "unknown for *I.*

pliacus" (1968, p. 278), as it was preserved on some of the skull material that I referred (Wood, 1937, p. 190) to *I. pliacus*, although, not surprisingly, it was not preserved on the holotype lower jaw.

Because cheek-teeth are frequently the only parts of mammalian fossils either available for study or studied (Romer, 1969, p. 43n), the apparent lack of cheek-tooth differentiation among ischyromyids may make the identification of fragmentary fossils more difficult. However, the absence of such differentiation cannot be used to override the clear differences in skull structure, which imply major differences in jaw musculature and, presumably, in jaw mechanics and gnawing habits. Nor should the difficulty or impossibility of assigning isolated cheek-teeth or lower jaws of ischyromyids to the correct genus, without association of at least skull fragments, be considered any more of a real problem than the comparable difficulty with respect to isolated caudal vertebrae or toe bones.

Taxonomic Relationships

The data presented above show that *Titanotheriomys* is extremely different in its masseteric and other cephalic musculature from any paramyid, that *Ischyromys* is clearly intermediate in these respects between paramyids and *Titanotheriomys*, and that all paramyids (except *Manitsha*) are very uniform in the patterns of the Mm. masseteres superficiales and laterales. The positions and relationships of numerous cranial foramina and the presence or absence of several others show that there had been considerable changes in the innervation and vascularization of the head; at least some of these must have been independent of the changes in the musculature.

The presence of uniserial incisive enamel in the ischyromyids is another significant difference from the pauciserial paramyids. Only one case of change from pauciserial to uniserial or multiserial enamel has as yet been demonstrated – that in the Theridomyidae where *Theridomys* and *Protechimys* are pauciserial, *Archaeomys* is uniserial, and *Issiodoromys* (*Nesokerodon*) shows a transitional condition (Korvenkontio, 1934, p. 116). It should be noted that Stehlin and Schaub (1951, pp. 82–83) place *Protechimys* in the synonymy of *Archaeomys*. I do not know whether this means that some species of *Archaeomys* were pauciserial and some uniserial, or that some of Korvenkontio's specimens were misidentified. Obviously, theridomyid incisor enamel needs further investigation.

Therefore, I believe that *Titanotheriomys* and *Ischyromys* are distinct, valid genera; that *Titanotheriomys* deserves familial separation from the Paramyidae; and that *Ischyromys* clearly is closer to *Titanotheriomys* than to any known paramyid and belongs in the same family as *Titanotheriomys*, despite the fact that, if jaw musculature were considered of subordinal importance, *Titanotheriomys* and *Ischyromys* would be placed in different suborders. Therefore, I continue to recognize the Ischyromyidae and Paramyidae as distinct families.

The forward migration of the origin of the M. masseter lateralis, off the ventral root of the zygoma, lateral and anterior to the infraorbital foramen (Figs. 3A, ML; 4C, ML; Wood, 1937, pl. 27, figs. 1, 1A), is of great importance in studies of rodent evolution (Wood, 1937, pp. 194–195). The discovery of the transitional forms from the primitive protrogomorphous condition of the Paramyidae to produce the sciuromorphous, myomorphous, hystricomorphous-sciurognathous and hystrico-morphous-hystricognathous types is of major importance in clarifying rodent phylogeny. The origins of any of these advanced types have rarely been demonstrated (see below).

Titanotheriomys had already achieved the transition from the protrogomorphous condition of paramyids, via the modified protrogomorphous condition of *Ischyromys*, to a sciuromorphous (possibly even myomorphous) condition (Table 3), although the forwardly displaced part of the M. masseter lateralis was clearly much less massive than in modern sciuromorphous rodents.

An enlargement of the infraorbital foramen and a concomitant forward movement of the origin of the M. masseter medialis can be followed in the European Eocene-Oligocene series *Protadelomys cartieri* (Lutetian) – *Sciuroides* sp. (Ludian: Hartenberger, 1969) – *Adelomys* sp. (late Eocene); even in the first of this series, the M. masseter medialis had penetrated the somewhat enlarged infraorbital foramen and had reached almost to the premaxillary-maxillary suture (Wood, 1974b, fig. 4).

Intergeneric differences in contemporary populations, similar to the differences between *Titanotheriomys* and *Ischyromys*, can be seen in the early Oligocene Deseadan fauna of South America, where *Platypittamys* has a small infraorbital foramen, as in *Protadelomys cartieri*, and no visible scar for the M. masseter medialis on the snout in the front of the foramen; *Cephalomys* and *Branisamys* have very large infraorbital foramina, with the origin of the M. masseter medialis reaching almost to the tip of the snout; and *Incamys* has an intermediate condition (Wood, 1949, fig. 2A; Wood and Patterson, 1959, fig. 21C; Patterson and Wood, in preparation).

Black (1963, pp. 135 and 138) stated that the early sciurids *Miosciurus* and *Protosciurus* showed no forward extension of the M. masseter lateralis above the infraorbital foramen; he gave no illustrations of this area in *Miosciurus*; his figure of *Protosciurus rachelae* (pl. 6, fig. 1) suggests that here the muscle had already extended dorsal and anterior to the infraorbital foramen, but that it was smaller than in later sciurids.

Hartenberger (1971, p. 112) stated that the skull of *Gliravus majori* from the middle Oligocene of Quercy shows that "celui-ci est manifestement une forme protrogomorphe." His phylogenetic chart (fig. 16) shows *Gliravus* as ancestral to the modern glirids. Therefore, although the transitional stages have not yet been demonstrated, there must have been modification of the jaw musculature within the Gliridae, independent of changes in any other rodents. However, the fact that the living European and Asiatic genera are myomorphous whereas the African forms are hystricomorphous (Tullberg, 1899, p. 172 and pl. 11, figs. 8–10, 18–19) leaves the question whether the change from protrogomorphy was to myomorphy and then to hystricomorphy or the reverse.

Finally, the M. masseter medialis of the Bathyergidae may be regressing from an ancestral hystricomorphous type. Such a change has only recently been demonstrated (Lavocat, 1973, pp. 151–152).

Family Ischyromyidae Alston, 1876

Revised Diagnosis

Rodents with a zygomatic arch elevated on the side of the skull, increasing the length of the fibres of the masseter; M. masseter lateralis shifted dorsally at its anterior end and, in *Titanotheriomys* its origin spread anterodorsally, lateral to the infraorbital foramen, reaching at least to the premaxillary-maxillary suture near the dorsum of the snout; origin of the M. masseter superficialis separated by 2–3 mm from that of the M. masseter lateralis; M. temporalis of variable size, tending to be smaller in those forms in which the M. masseter lateralis had moved forward; snout musculature better differentiated than in paramyids; compared with paramyids, orbital foramina displaced anteriorly relative to cheek-teeth; orbitosphenoid reaches sphenopalatine foramen; sphenofrontal foramen reduced or absent; foramen between maxilla and palatine behind M^3; V^3 foramen through ectopterygoid in front of bulla; masticatory and buccinator foramina usually united with V^3 foramen; large septate bulla co-ossified with skull; premaxillary-maxillary suture nearly vertical on side of snout; posterior ends of nasals and premaxillae generally a nearly straight line, slightly anterior to the front of the orbits; lower incisors ovate in cross-section, with narrow end forward; incisive enamel uniserial; cheek-teeth brachydont and four-crested, with little or no trace of separate cusps, except in primitive species; skeleton generalized, primitive, adapted to scampering to sub-fossorial habits.

Referred Genera

Ischyromys and *Titanotheriomys*.

Distribution

Oligocene of North America; isolated teeth are referred to the family from the latest Eocene of Badwater, Wyoming.

Genus *Ischyromys* Leidy, 1856

Revised Diagnosis

Ischyromyid with primitive masseter; origin of the M. masseter lateralis entirely posteroventral to infraorbital foramen; origin of M. masseter superficialis relatively close to that of M. masseter lateralis; M. temporalis always large; sagittal crest in adults half or more the length of the skull; facial musculature not well developed.

Distribution

Oligocene of North America.

Genus *Titanotheriomys* Matthew, 1910

Revised Diagnosis

Ischyromyid with advanced, sciuromorphous (or myomorphous?) masseter; anterior part of origin of M. masseter lateralis migrated forward lateral to and above infraorbital foramen, originating from as far forward as the premaxillary-maxillary suture, its fibres nearly parallel to those of M. masseter superficialis; origin of M. masseter superficialis farther from that of M. masseter lateralis than in *Ischyromys* and more rugose; M. temporalis often reduced in size with lyrate muscle-free area on the dorsum of skull; sagittal crest very short or absent; facial musculature well developed.

Distribution
Early and middle Oligocene of North America.

Distribution of Ischyromyidae

Previously I considered *Titanotheriomys* to be early Oligocene and *Ischyromys* primarily middle and late Oligocene (Wood, 1937). Since in my opinion *Titanotheriomys* is more advanced than *Ischyromys*, it would be surprising if it antedated the latter. Black (1968) reduced the previously described species to two, *I. typus* restricted to middle and possibly late Oligocene, and *I. veterior* ranging from early through middle Oligocene, its temporal and geographic ranges overlapping those of *I. typus*. This raises the problem of sympatry, especially if the two forms were as closely related as Black implied. These difficulties are now, I believe, partially eliminated, the first by the discovery of early Chadronian specimens of *Ischyromys*, and the second by the recognition of *Titanotheriomys* as a distinct genus.

Vieja Ischyromyids
A partial ischyromyid skull (TMM 41216-10) was found in the Porvenir local fauna of the Chambers Tuff, slightly younger than the Buckshot Ignimbrite (38.6 m.y.; Wilson *et al.*, 1968, p. 599). This is presumably the oldest known ischyromyid skull, as the oldest published date for the base of the Bates Hole sequence (Evernden *et al.*, 1964, p. 185, KA 895) is 35.2 m.y., and the Porvenir local fauna is clearly considerably older than that of Pipestone Springs and appreciably older than that of McCarty's Mountain (Wood, 1974a, p. 105).

The Porvenir skull bears a pronounced fossa on the ventral surface of the zygoma for the M. masseter lateralis, with its anterior end marked as in the typical *Ischyromys*. There is a very rugose area, close to the median end of the M. masseter lateralis fossa, clearly for the attachment of the tendon of the M. masseter superficialis. The position of this rugose area is more like that in typical paramyids, sciuravids or cylindrodonts than are those in later ischyromyids. The pit for the pars maxillaris anterior of the M. buccinator is shallow and small. The lateral surface of the premaxilla is damaged, but no fossa for the M. bucco-naso-labialis was noted. The supraorbital crests bend abruptly medially and form two closely parallel temporal crests.

Since this specimen is juvenile, retaining deciduous premolars although M^3 had erupted, the double temporal crests presumably were the juvenile condition and, had the individual lived to be an adult, they would have fused as in typical *Ischyromys*. This skull, the earliest known for the family and a juvenile, is the only ischyromyid that I know with the *Ischyromys*-type of M. masseter lateralis and an incomplete sagittal crest. The cheek-teeth are in about the same stage of evolution as those of *T. douglassi* from McCarty's Mountain, and I have described the Porvenir form as *Ischyromys blacki*, n. sp. (Wood, 1974a, p. 22).

Ischyromyid specimens are abundant in the Ash Spring local fauna of the Vieja Group, which is clearly Chadronian, and probably the latest local fauna of the Vieja, although there are no controlling stratigraphic or K/A correlations. A damaged skull (TMM 40289-91) and a very fragmentary specimen including both lower jaws, the premaxillae and part of the nasals, and the left and some right upper cheek-teeth (TMM 40289-126), permit the masseteric condition to be determined. The origin of the M. masseter lateralis clearly did not have its front boundary at the anterior end of the cheek-teeth, but extended forward lateral to the infraorbital foramen, although damage to the snout makes it impossible to determine precise limits. The dorsal part of the skull is badly damaged, but there is no trace of a sagittal crest in the interparietal region, and the temporal ridges must have been widely separated, as they are not preserved. There are suggestions in the postorbital region that the crests were lyrate.

There is a small fossa, lateral to P^3, for the tendon of origin of the M. masseter superficialis. This is, for an ischyromyid (and particularly for *Titanotheriomys*), close to the fossa for the M. masseter lateralis but completely separate from it. Posterolateral to, and partially closed ventrally by, the diastemal ridge is a deep pit for the origin of the pars maxillaris anterior of the M. buccinator. The cheek-teeth of the Ash Spring ischyromyids are distinctly advanced over those of the Porvenir species, and are indistinguishable from teeth of *Titanotheriomys veterior*.

McCarty's Mountain Ischyromyids

A series of five skulls from McCarty's Mountain, Montana, in the Carnegie Museum collections (CM 1053, 1122, 1123, 10966, and 10967) and three in the collections of Montana State University at Missoula (UM 0740, 0929 and 11051) belong to the ischyromyid that Black (1968, p. 283) described as *Ischyromys douglassi*. This species is very primitive in the cheek-tooth pattern, as indicated by Black's diagnosis. The teeth are crested, but all the cusps are clearly evident, in contrast to all other ischyromyids, except the Porvenir *Ischyromys* discussed above.

Sagittal crests are present on all five Carnegie Museum specimens; this area is not preserved on the Montana State University specimens. The zygoma and the oblique plate below the infraorbital foramen are damaged in some specimens. In CM 1122 the M. masseter lateralis clearly occupied all the lateral two-thirds of the plate, but may not have reached the ventromedial corner of the infraorbital foramen. It is not clear whether this last is due to post-mortem damage. The fossa of the M. masseter lateralis extends lateral to the infraorbital foramen, perhaps as far as the posterior part of the premaxilla. Essentially the same conditions are present on CM 1053 and 1123 (Fig. 5A). Damage to the masseteric plates and zygomata of all other specimens prevents identification of the anterior margin of the M. masseter lateralis. There is no evidence of the medial curve of the anterior end of this muscle, characteristic of *Ischyromys*.

The fossa for the M. bucco-naso-labialis is unusual in that it extends from the premaxilla back onto the maxilla; in CM 1122 and 1123 it looks as though it carried a muscle that passed through the infraorbital foramen, as in AMNH F:AM 79312 from Bates Hole (Fig. 7; X). A horizontal crest crosses the median wall of the orbit, above the frontomaxillary suture and below the maxillolacrimal one, leading to the lower third of the infraorbital foramen. This suggests that the upper two-thirds of the foramen may have been occupied by something other than the infraorbital nerves and blood vessels.

The pits for the M. masseter superficialis and the pars maxillaris anterior of the M. buccinator vary in depth and rugosity. They are preserved in CM 1053, 1122 and 1123; this area is damaged in CM 10966.

If the structure of the M. masseter lateralis is the diagnostic feature of *Titano-theriomys*, these specimens belong to that genus. If *Titanotheriomys* is descended from *Ischyromys*, as seems probable, *I. douglassi* is primitive both by its posses-sion of a sagittal crest and by its tooth structure. This is reasonable, as McCarty's Mountain is probably rather early Chadronian. Since I believe that the pattern of the M. masseter lateralis is taxonomically diagnostic in ischyromyids, this species should be transferred to *Titanotheriomys*.

Bates Hole Ischyromyids

A series of 14 skulls of ischyromyids from Bates Hole, Wyoming, in the Frick Collections of the American Museum, were collected between 35 feet ($= 11$ m) below and 20 feet ($= 6$ m) above Ash D (285-foot ash). This ash lies about half-way between Ash B and Ash G, for which Evernden *et al.* (1964, p. 185) reported K/A dates of 33.3 (or 35.2) and 32.6 m.y. respectively. These skulls all show the highly advanced zygomasseteric pattern characteristic of *Titanotheriomys*, usually associated with a lyrate area between the temporal muscles.

Each of these skulls has a depression on the maxilla between the alveolus of P^3 and the infraorbital foramen (Fig. 6A, B; MS) that seems to mark the attachment of the tendon of origin of the M. masseter superficialis. Certainly there is no other area for the origin of this muscle. The depressions are highly variable in size, in the strength of their borders, and in rugosity.

Directly in front of P^3 and just lateral to the diastemal ridge, a deep fossa faces ventrolaterally. It may be a hemispherical depression (Fig. 6A; BU), or it may be bounded ventromedially by a strong shelf (Fig. 7; BU) so that it opens directly laterad. This is the area of origin of the pars maxillaris anterior of the M. buc-cinator.

Of the nine skulls that preserve the dorsal surface, two (AMNH F:AM 79313 and F:AM 79318) have sagittal crests and seven (AMNH 42993, 42995, F:AM 79312 and F:AM 79314–F:AM 79317) show lyrate intermuscular areas of various sizes.

On one of the skulls a ridge (Fig. 7; X) runs from the infraorbital foramen to the premaxillary fossa assigned above to the M. bucco-naso-labialis (Fig. 7; BNL), suggesting that a slip of the M. masseter medialis possibly penetrated the infra-orbital foramen.

The Bates Hole ischyromyid skulls clearly demonstrate that the M. masseter lateralis had migrated forward, lateral to the infraorbital foramen, onto the snout; that the origin of the M. masseter superficialis was farther from that of the M. masseter lateralis than in paramyids; and that there may have been a forward migration of a slip of the M. masseter medialis through the infraorbital foramen to the premaxilla. Associated with these changes, M. temporalis usually was reduced in size.

A zone of red-weathering clays near the bottom of the section in Bates Hole is identified on field labels as the "Red Fauna Zone". This is about 50 feet (= 16 m) below Ash B, and 75 feet (= 24 m) below the level of the lowest of the 14 skulls listed above. A rather poorly preserved skull from this level (AMNH 43000) shows that the masseter was very different from that described above, and apparently was identical with the condition typical of *Ischyromys*. The skull has strong temporal ridges.

Pipestone Springs and Beaver Divide Ischyromyids

Four skulls or partial ischyromyid skulls in the Carnegie Museum (CM 8924, 9058, 10660 and 17453) and one in my collection (AEW 9048) from Pipestone Springs, Montana, preserve at least the snout and the anterior root of the zygomatic arch. This last specimen was part of my 1937 hypodigm of *Titanotheriomys*, contrary to Black (1968, p. 279). The pattern of scars is that characteristic of *Titanotheriomys*, with a fossa for the M. masseter lateralis extending laterally from the infraorbital foramen forward to the premaxillary-maxillary suture, a clearly marked premaxillary fossa extending anterior to the upper incisor and almost to the external nares, and a prominent pit in front of P^3 for the pars maxillaris anterior of the M. buccinator. The depression for the M. masseter superficialis is shallow on all specimens except CM 9058, in which it is very deep. When AEW 9048 is held at the correct angle, reflected light shows that the posterodorsal boundary of the premaxillary fossa extends across the maxilla toward the infraorbital foramen, as in AMNH F:AM 79312 from Bates Hole. Only CM 9058 has a sagittal crest; the others show variations of the lyrate intermuscular area on top of the skull.

The Beaver Divide skulls, referred by Black (1968) to *I. veterior*, have lyrate temporal crests, a forwardly displaced M. masseter lateralis, and a very pronounced pit for the pars maxillaris anterior of the M. buccinator (Wood, 1937, p. 198).

Cypress Hills Ischyromyids

Russell (1972, pp. 27–30) discussed the ischyromyids from the Cypress Hills of Saskatchewan. These had previously been described as *I. typus* (Lambe, 1908, p. 56), *I. typus nanus* (Russell, 1934, p. 56) and *I. parvidens* (Wood, 1937, p. 192). Russell (1972) described, from additional material, a new species of *Ischyromys*, *I. junctus*, separated by characters detectable in isolated cheek-teeth from both *Titanotheriomys* and other species of *Ischyromys*. He referred small upper cheek-teeth to *Ischyromys* sp. Unfortunately, no diagnostic cranial material is available.

Chadronian Ischyromyids from Douglas, Wyoming

A series of ischyromyid skulls from southeast of Douglas, Wyoming, in the Frick Collection in the American Museum, came from the nodular zone, which underlies a stream channel from which a titanothere jaw and limb fragments were collected and is therefore presumably Chadronian (oral comm., Morris F. Skinner). The stratigraphic and faunal relationships, however, suggest that it is considerably later than the Bates Hole deposits.

In these specimens, there is a diagonal plate below and behind the infraorbital foramen, but the fossa for the M. masseter lateralis is clearly limited to the posterior part of this plate. The anterior margin of the fossa is deeply entrenched along the zygoma lateral to P^3–M^1, and then curves mediad across the middle of the diagonal plate, halfway to P^3, before fading out.

The fossae for the M. masseter superficialis and the pars maxillaris of the M. buccinator both have highly rugose surfaces. The former is the larger, lies on the diagonal plate between the alveolus of P^3 and the infraorbital foramen, and is larger and more rugose than those in the Bates Hole specimens.

None of the material from Douglas shows any trace of muscle attachment lateral to the infraorbital foramen or on the maxilla, in front of the foramen. The fossa for the M. bucco-naso-labialis is poorly developed. All specimens have well-developed sagittal crests.

Orellan Ischyromyids

The characteristic zygoma of middle Oligocene ischyromyids shows a fossa for the M. masseter lateralis similar to that in the Douglas specimens, but extending slightly closer to the midline. The fossa for the M. masseter superficialis is generally rather small, and that for the pars maxillaris anterior of the M. buccinator is very small. There are well-developed sagittal crests.

Occasionally, however, middle Oligocene ischyromyids exhibit a different condition as seen in a specimen from Slim Buttes, South Dakota (Burt and Wood, 1960, fig. 1B). The fossa for the M. masseter lateralis is normal for *Ischyromys*. There is a deep fossa for the M. masseter superficialis that occupies most of the zygomatic plate between P^3 and the infraorbital foramen; the pit in front of P^3 for the pars maxillaris anterior of the buccinator is broad but shallow, and without rugosities. Burt and Wood (1960, p. 968) referred this specimen to *I. troxelli*, which was originally defined partly on this type of zygomasseteric pattern (Wood, 1937, p. 191). This species may be separable from *I. typus* on cranial characteristics, but no adequate study of cranial variation in *Ischyromys* has yet been made.

Almost without exception middle Oligocene ischyromyids have been referred to *Ischyromys* and have been considered specifically distinct from the Chadronian form. Black (1968, p. 283), however, referred specimens previously identified as *Ischyromys parvidens* from the middle Oligocene to *Ischyromys* (= *Titanotheriomys*) *veterior*, on the basis of statistical analyses of cheek-teeth. I have no information on the jaw musculature of those specimens as no skull material of *I. parvidens* was reported by me (1937) or Black (1968). I have listed them on the distribution chart (Fig. 9) as "*T. veterior* or *parvidens*", assuming them to have had the *Titanotheriomys*-type of jaw musculature.

Conclusions

The geographic and geologic distribution of the Ischyromyidae, as presently understood, is summarized in Figure 9. If I am correct that *Ischyromys* and *Titanotheriomys* are distinct genera, separable at least on the basis of the M. masseter lateralis,
Ischyromys (the more primitive) appears first in the Chambers Tuff, the earliest
known Oligocene deposit that has produced ischyromyids. The genus is relatively
rare in the Chadronian, but occurs at the base of the Bates Hole sequence, apparently in the Cypress Hills, and in the late Chadronian of Douglas, Wyoming; it is
dominant in the Orellan. I suspect that Black's (1971, pp. 203–204) *Ischyromys*
sp. from the late Eocene of Badwater, Wyoming, will prove to be *Ischyromys?*,
when diagnostic parts are found. *Titanotheriomys*, distinctly more advanced in the
structure of the M. masseter lateralis, is represented by a primitive species at
McCarty's Mountain and by *T. veterior* at Pipestone Springs, Beaver Divide, at
all but the base of the Bates Hole sequence, and in the Ash Spring local fauna of
the Vieja. A relatively small population survived into the Orellan.

Initially, in the late Eocene, there was probably rapid progress in the development of crested cheek-teeth, uniserial enamel on the incisors, and the type of
zygomasseteric structure seen in *Ischyromys*, which is associated with a large M.
temporalis. In one lineage (*Ischyromys*), the M. masseter remained on the zygoma

WHITNEYAN	*I. typus*	
ORELLAN	*I. typus* (abundant)	*T. veterior* (or *parvidens*)
CHADRONIAN	*I.* cf. *typus* Douglas, Wyo. *I.* cf. *typus* Bates Hole, Red Zone *I. junctus* Cypress Hills, Sask. *Ischyromys* n. sp. Porvenir L.F., Vieja	*T.* cf. *veterior* Bates Hole, Wyo. *T. veterior* Ash Spring L.F., Vieja *T.* cf. *veterior* Beaver Divide, Wyo. *T. veterior* Pipestone Springs, Mont. *T. douglassi* McCarty's Mt., Mont.

Fig. 9 Distribution of Oligocene ischyromyids.

and the M. temporalis remained large. There seems to have been minimal evolution during most of the Oligocene—at most a slight change in the average size of the cheek-teeth (Black, 1968, pp. 290–300) and possibly minor changes in the jaw musculature. In the other lineage, the M. masseter lateralis moved forward off the zygoma and established its origin on the snout at least as far as the anterior end of the maxilla. *T. douglassi*, the earliest known member of this lineage, retained an unreduced M. temporalis and hence a long sagittal crest. Subsequently, the size of the Mm. temporales was reduced, leaving a space between the muscles of the two sides and forming lyrate supraorbital crests. Some specimens suggest that the M. masseter medialis was also involved in the changes. The facial musculature was stronger than in *Ischyromys*. *Titanotheriomys* may have survived into the Orellan as the forms previously identified as *Ischyromys parvidens*.

Among known Eocene rodents, some paramyids and *Protoptychus* (Wilson, 1937, fig. 1) show the greatest similarities in cheek-tooth pattern to the Ischyromyidae. The arrangement of the cranial foramina (a notoriously conservative feature) in the Ischyromyidae is most like that of the Cylindrodontidae, which differ markedly from the ischyromyids in tooth pattern. The very similar tooth pattern of *Protoptychus* is accompanied by a highly inflated auditory bulla (saltatory adaptation?) and a M. masseter medialis that had invaded the infraorbital foramen (Wahlert, 1973). These are associated with a M. masseter lateralis limited to the ventral margin of the zygoma (a hystricomorphous combination) and, apparently, a hystricognathous angular process (Wahlert, 1973).

The similarities between ischyromyids and *Leptotomus* pointed out above and by Black (1968, pp. 298–299) seem more significant than those with *Rapamys* (see Black, 1968) because *Rapamys* is sub-hystricognathous. Although I believe that ischyromyids and *Leptotomus* are related, I do not feel as certain of this as I once did. The suggested invasion of the infraorbital foramen by the M. masseter medialis in both *Protoptychus* and *Titanotheriomys* suggests relationships that should be further investigated (even though the M. masseter lateralis of the latter has also shifted forward onto the snout, so that perhaps *Titanotheriomys* was myomorphous).

Whether *Titanotheriomys* was sciuromorphous or myomorphous, it clearly had no relationships with other sciuromorphous or myomorphous rodents, and seems to have left no descendants.

Acknowledgments

The availability in the Frick Collection of the American Museum of a series of ischyromyid skulls from the Chadronian of Bates Hole and Douglas, Wyoming, has greatly facilitated the development of the ideas expressed, and I am grateful to Dr. Malcolm C. McKenna for permission to study these specimens. Mr. Morris F. Skinner provided invaluable assistance with his encyclopaedic knowledge of the localities and levels where these specimens were found. Dr. Robert J. Emry checked details on some of the skulls for me. I have profited greatly from discussions with Drs. Craig C. Black, Mary Dawson, and Richard H. Tedford. With their extensive knowledge of rodent myology, Dr. David J. Klingener of the University

of Massachusetts and Dr. Charles A. Woods of the University of Vermont have
guided me in the interpretation of the musculature of the ischyromyids. The manuscript has been greatly improved by the comments of Drs. C.S. Churcher and T.S.
Parsons. Any errors which persist in spite of all this help are due to my own
stubbornness in clinging to my original ideas. This study was assisted by Grant
GB-6075 from the National Science Foundation, Grants 143 and 145 from the
Marsh Fund of the National Academy of Sciences, and financial assistance from
the Biology Department, Amherst College.

Literature Cited

ALSTON, E.R.
   1876   On the classification of the order Glires. Proc. Zool. Soc. London, 1876(1): 61–98.

BLACK, C.C.
   1963   A review of the North American Tertiary Sciuridae. Bull. Mus. Comp. Zool. Harv.,
           130(3): 109–248, figs. 1–8, pls. 1–22.
   1968   The Oligocene rodent *Ischyromys* and discussion of the family Ischyromyidae.
           Ann. Carneg. Mus., 39(art. 18): 273–305, figs. 1–26.
   1971   Paleontology and geology of the Badwater Creek area, central Wyoming. Part 7.
           Rodents of the family Ischyromyidae. Ann. Carneg. Mus., 43(art. 6): 179–217,
           figs. 1–73.

BRANDT, J.F.
   1855   Untersuchungen über die craniologischen Entwicklungstufen und die davon herzu-
           leitenden Verwandtschaften und Classification der Nager der Jetzwelt. Mém. Acad.
           Imp. St. Pétersbourg, ser. 6, 7: 127–336.

BURT, A.M. AND A.E. WOOD
   1960   Variants among Middle Oligocene rodents and lagomorphs. J. Paleont., 34(5):
           957–960, figs. 1–2.

EVERNDEN, J.F., D.E. SAVAGE, G.H. CURTIS AND G.T. JAMES
   1964   Potassium-argon dates and the Cenozoic mammalian chronology of North
           America. Am. J. Sci., 262: 145–198.

HARTENBERGER, J.-L.
   1969   Les Pseudosciuridae (Mammalia, Rodentia) de l'Eocène moyen de Bouxwiller,
           Egerkingen et Lissieu. Palaeovertebrata, 3(2): 27–61, figs. 1–6, pls. 1–4.
   1971   Contribution à l'étude des genres *Gliravus* et *Microparamys* (Rodentia) de
           l'Eocène d'Europe. Palæovertebrata, 4: 97–135.

HOWE, J.A.
   1966   The Oligocene rodent *Ischyromys* in Nebraska. J. Paleont., 40(5): 1200–1210,
           figs. 1–3.

KORVENKONTIO, V.A.
   1934   Mikroskopische Untersuchungen an Nagerincisiven unter Hinweis auf die Schmelz-
           struktur der Backenzähne. Histologisch-phyletische Studie. Ann. Zool., Soc. Zool.-
           Bot. Fennicae Vanamo, 2(1): i–xiv, 1–274, figs. 1–15, pls. 1–47 and 3 supple-
           mentary figs.

LAMBE, L.M.
   1908   The Vertebrata of the Oligocene of the Cypress Hills, Saskatchewan. Contr. Can.
           Paleont., 3(4): 1–64, figs. 1–13, pls. 1–8.

LAVOCAT, R.
  1973   Les Rongeurs du Miocène d'Afrique Orientale. 1. Miocène Inférieur. Ecole
          Pratique des Hautes Etudes; Institut de Montpelier, Mém 1: i–v, 1–284, figs. 1–20,
          maps 1–2, and 44 pls. in a separate folder.

LEIDY, J.
  1856   Notice of remains of extinct Mammalia, discovered by Dr. F.V. Hayden in
          Nebraska Territory. Proc. Acad. Nat. Sci. Philad., 8(2): 88–90.

MATTHEW, W.D.
  1910   On the osteology and relationships of *Paramys*, and the affinities of the Ischyro-
          myidae. Bull. Am. Mus. Nat. Hist., 28: 43–71, figs. 1–19.

MEINERTZ, T.
  1943   Das superfizielle Facialisgebiet der Nager. VI. Die Sciuriden. 2. *Marmota marmota*
          (L). Zeit. f. Anat. u. Entwicklungsgeschichte, 112: 350–381.

MILLER, G.S. AND J.W. GIDLEY
  1920   A new fossil rodent from the Oligocene of South Dakota. J. Mammal., 1(2):
          73–74.

PATTERSON, B. AND A.E. WOOD
  in prep. The caviomorph rodents from the early Oligocene of Bolivia and their place in
          caviomorph evolution.

ROMER, A.S.
  1969   Teaching vertebrate paleontology. Proc. North. Am. Paleont. Convention, Part A:
          39–46.

RUSSELL, L.S.
  1934   Revision of the Lower Oligocene vertebrate fauna of the Cypress Hills, Saskatch-
          ewan. Trans. R. Can. Inst., 20(pt. 1, no. 43): 49–67, pls. 7–10.
  1972   Tertiary mammals of Saskatchewan Part II: The Oligocene fauna, non-ungulate
          orders. Life Sci. Contr., R. Ont. Mus., 84: 1–97, figs. 1–17.

STEHLIN, H.G. AND S. SCHAUB
  1951   Die Trigonodontie der simplicidentaten Nager. Schweizerische paläontologische
          Abhandl., 67: 1–385, figs. 1–620.

TROXELL, E.L.
  1922   Oligocene rodents of the genus *Ischyromys*. Am. J. Sci., ser. 5, 3: 123–130, figs.
          1–7.

TULLBERG, T.
  1899   Ueber das System der Nagethiere, eine phylogenetische Studie. Nova Acta Reg.
          Soc. Scient., Upsala, Ser. 3, 18: 1–514, pls. 1–56.

WAHLERT, J.H.
  1968   Variability of rodent incisor enamel as viewed in thin section, and the microstruc-
          ture of the enamel in fossil and recent rodent groups. Breviora Mus. Comp. Zool.
          Harv., 309: 1–18, figs. 1–3.
  1973   *Protoptychus*, a hystricomorphous rodent from the late Eocene of North America.
          Breviora Mus. Comp. Zool. Harv., 419: 1–14, figs. 1–2.
  1974   The cranial foramina of protrogomorphous rodents: an anatomical and phylo-
          genetic study. Bull. Mus. Comp. Zool. Harv., 146(8): 363–410 figs. 1–13.

WATERHOUSE, G.R.
  1839   Observations on the Rodentia, with a view to point out the groups, as indicated by
          the structure of the crania, in this order of mammals. Mag. Nat. Hist., n.s., 3:
          90–96.

WILSON, J.A., P.C. TWISS, R.K. DE FORD AND S.E. CLABAUGH

1968   Stratigraphic succession, potassium-argon dates, and vertebrate faunas, Vieja Group, Rim Rock Country, Trans-Pecos Texas. Amer. J. Sci., 266: 590–604, figs. 1–2.

WILSON, R.W.

1937   Two new Eocene rodents from the Green River Basin, Wyoming. Am. J. Sci., ser. 5, 34: 447–456, figs. 1–2.

1940   Two new Eocene rodents from California. Carneg. Inst. Washington Publ., 514(6): 85–95, pls. 1–2.

WOOD, A.E.

1937   Part II. Rodentia. *In* Scott, W.B., G.L. Jepsen and A.E. Wood. The mammalian fauna of the White River Oligocene. Trans. Am. Phil. Soc., n.s., 28(pt. 2): 155–269, figs. 8–70, pls. 23–33.

1940   Part III. Lagomorpha. *In* Scott, W.B., G.L. Jepsen and A.E. Wood. The mammalian fauna of the White River Oligocene. Trans. Am. Phil. Soc., n.s., 28(pt. 3): 271–362, figs. 71–116, pls. 34–35.

1949   A new Oligocene rodent genus from Patagonia. Am. Mus. Nat. Hist., Novitates 1435: 1–54, figs. 1–8.

1962   The early Tertiary rodents of the family Paramyidae. Trans. Am. Phil. Soc., n.s., 52: 1–261, figs. 1–91.

1970   The European Eocene paramyid rodent, *Plesiarctomys*. Verhandl. Naturf. Gesell. Basel, 80(2): 237–278, figs. 1–15.

1974a   Early Tertiary vertebrate faunas, Vieja Group, Trans-Pecos Texas: Rodentia. Texas Memorial Mus., Bull. 21: 1–112, figs. 1–40.

1974b   The evolution of the old world and new world Hystricomorphs, *in* The biology of hystricomorph rodents: I.W. Rowlands and B.S. Weir, eds. Symp. Zool. Soc. London, 34: 21–60, figs. 1–6.

WOOD, A.E. AND B. PATTERSON

1959   The rodents of the Deseadan Oligocene of Patagonia and the beginnings of South American rodent evolution. Bull. Mus. Comp. Zool. Harv., 120(3): 281–428, figs. 1–35.

The Localities of the Cudahy Fauna, with a New Ground Squirrel (Rodentia, Sciuridae) from the Fauna of Kansas (Late Kansan)

Claude W. Hibbard
Late Professor of Geology, Museum of Paleontology,
The University of Michigan, Ann Arbor

Abstract

Previous placement of the late Kansan Cudahy fauna below the Aftonian Borchers warm local fauna in Meade County, Kansas, is explained. All white volcanic ashes in the Plains Region were once believed to be the Pearlette ash. The assignment of a Yarmouthian age to part of the Cudahy fauna was based on fossils from the Sappa Formation of Nebraska. A ground squirrel, *Spermophilus lorisrusselli*, is described from the Cudahy fauna (Wilson faunule) of Lincoln County, Kansas. The specimen was taken just below the "Pearlette"-like, type O, volcanic ash (600,000 yrs.).

Introduction

The late Kansan Cudahy fauna (Hibbard, 1944) was originally described from vertebrates taken from silts directly below the "Pearlette" ash. The silts and ash were assigned to part of the Meade Formation. I stated (p. 752), "The fauna indicates a cooler and more humid climate than now exists in that region. It is thought to have lived in that area at the close of a glacial stage." The fauna was first collected by A. Byron Leonard and George C. Rinker in 1941 who took matrix from below the "Pearlette" ash in Meade County, Kansas, and washed it to recover molluscs. It was also found to contain vertebrates.

Frye *et al.* (1948, p. 506) give the age of the volcanic ash and the underlying molluscan fauna as Yarmouthian. Leonard (1950) describes the Yarmouthian molluscan fauna taken in the midcontinent region. Leonard states that "Eighteen of the localities of the molluscan fauna are intimately associated with the Pearlette ash." Some of these localities include the Cudahy fauna in Kansas.

Moore *et al.* (1951, p. 14) stated, "A fall of petrographically distinct volcanic ash occurred in latest Kansan time." This applied to the volcanic ashes in the Pleistocene of Kansas. The "Pearlette" volcanic ash bed has been said to occur in the Sappa Member of the Meade Formation (Moore *et al.*, 1951, fig. 6; Frye and Leonard, 1965, p. 206, fig. 3).

Leonard (1952, p. 8) lists the molluscan assemblages of the Sappa silts in Kansas that occurred in late Kansan and early Yarmouthian times. These include the molluscan assemblages taken with the Cudahy fauna in Kansas.

Frye and Leonard (1952, pp. 158–159) assigned a late Kansan age to these molluscan faunas, some of which are associated with the Cudahy fauna. The "Pearlette" volcanic ash directly above the Cudahy fauna of late Kansan age in Meade County was shown by Naeser *et al.* (1971) to be 600,000 years old. To this Pearlette-like ash they assigned the name type O. They also reported the age of the Pearlette-like Borchers ash (type B) in the type section of the Crooked Creek Formation (Hibbard, 1949, p. 70) as 1.9 ± 0.2 m.y. The type section of this formation is in the N½ Sec. 21, and s½ Sec. 16, Township 33 s., Range 28 w., on the east side of Crooked Creek, Meade County, Kansas. Only the Borchers (type B) volcanic ash occurs in this section. Matrix has been washed below this ash but no fauna was recovered. A hiatus exists between the Crooked Creek Formation and the deposits containing the Cudahy fauna in Kansas.

Invertebrates and vertebrates occurring in the Sappa Formation in Nebraska have been considered by Schultz *et al.* (1972, fig. 4) to be equivalent to the Cudahy fauna. Izett *et al.* (1971) dates the Pearlette-like ash from the type Sappa Formation as 1.2 m.y. Boellstorff (pers. comm., March 23, 1973) names the type s ash as the Coleridge ash and gives a date of 1.21 ± 0.5 m.y. This accounts for the assignment of the [warm] fauna below and just above the type s ash in Nebraska as being mainly Yarmouthian in age (Schultz *et al.*, 1972, fig. 4).

This has a direct bearing upon the age of the Cudahy fauna since only one "Pearlette" ash was considered to occur in the Pleistocene of Kansas. The Borchers warm local fauna occurs just above the Borchers ash. The fauna has never seemed to correspond with its placement *in* the Yarmouthian or *above* the Cudahy fauna. No fauna is known from the silts just below the Borchers volcanic ash. Hibbard (1948, p. 596; 1949, p. 69; 1956, p. 146; 1958a, p. 55; 1958b, p. 11; 1970, p. 402, *in* Flint, 1971, figs. 21–22) and Hibbard *et al.* (1965, p. 514) have shown that the Cudahy fauna of late Kansan age, below the Borchers local fauna, was assigned to a Yarmouthian age in the belief that the Pearlette was the only ash to occur in the Plains Region. The Borchers warm local fauna was assigned an Aftonian age and the Sappa Formation of Nebraska to the Nebraskan and Aftonian by Skinner *et al.* (1972, fig. 60).

Smith (1940, p. 119) questioned the age of the volcanic ash falls in the Pleistocene of the Plains Region and stated, "There is however, no *a priori* reason for assuming that there were so few eruptions, or that the conditions affecting the concentration and preservation of the ash from any one eruption were equally favorable over wide areas and in different drainage systems."

History of Collecting

The first fossils collected from a pit in which a volcanic ash was evident were from Meade County, Kansas (Smith, 1938). A.B. Leonard and George C. Rinker in the summer of 1941 removed fossiliferous matrix containing molluscs from the Sunbrite Volcanic Ash-Pit, south of Meade, Kansas, and washing revealed both molluscs and vertebrates. On August 26, 1941, John C. Frye and I located the exposure in Lincoln County that yielded the Wilson Valley faunule. In September, 1941, Frye, Leonard, and I visited Russell County, Kansas, and located the Tobin faunule from below the Pearlette-like volcanic ash (Frye *et al.*, 1943).

The summer of 1942 was spent obtaining fossiliferous matrix for washing from just below the Pearlette-like volcanic ash (type O, Pearlette, restricted; Boellstorff, 1973) in Meade County; two tons were taken in Clark County and one and one-half tons from the Tobin faunule site in Russell County. Hibbard (1944) described the Cudahy cool fauna including the Tobin and Wilson Valley faunules. The molluscs collected were reported by Frye *et al.* (1943, 1948), Leonard (1950), and Frye and Leonard (1952).

In the summer of 1958 fossiliferous matrix was collected from below the Pearlette volcanic ash (as restricted by Boellstorff, 1973; type O) at the type locality of the Cudahy fauna (KU Loc. 10, Meade County). A few specimens were also obtained from the Sunbrite Ash-Pit (KU Loc. 17). The vertebrates collected in 1958 are described by Paulson (1961).

Before 1962 Dalquest and his students from the Midwestern University, Wichita Falls, Texas, removed four tons of fossiliferous matrix from Baylor and Knox counties, from below the Pearlette-like ash, for washing and sorting. In the spring of 1962, William G. Melton, Jr., and I removed five tons of fossiliferous matrix from beneath the Pearlette-like ash at two localities in Knox County, Texas. This ash has never been dated and correlation is based on the fauna. This matrix was washed to recover microfossils. The molluscs obtained were described by Getz and Hibbard (1965) and the vertebrates by Hibbard and Dalquest (1966) as the Vera local faunule of the Cudahy fauna.

Hibbard and Dalquest considered the Seymour Formation in Knox County, Texas, to include the silts and clay containing the microfauna (Vera faunule), and the overlying Pearlette-like ash and silts. The assignment of these upper deposits to the Seymour Formation, *sensu stricto*, is in error since the Vera local faunule of the Cudahy fauna and the underlying deposits containing the Gilliland local fauna are separated by a long hiatus in deposition.

Localities

Locality 10 (Kansas University; Meade County, Kansas)

This locality is the Cudahy Ash-Pit in the sw¼ Sec. 2, T. 31 s., R. 28 w., type locality of the Cudahy fauna (Smith, 1938; Hibbard 1944; Frye *et al.* 1948, p. 516—Loc. 34; Leonard, 1950, p. 41—Loc. 34; Frye and Leonard, 1952, p. 159, located at SE¼ Sec. 2, T. 31 s., R. 28 w., Meade Co.; Paulson, 1961; Jammot, 1972).

Fig. 1 Localities of the Cudahy fauna.

Locality 17 (Kansas University; Meade County, Kansas)

This locality is the Sunbrite Ash-Pit, NW¼ SE¼ Sec. 26, T. 32 s., R. 28 w. Published reports on the Cudahy fauna from locality 17 are Hibbard (1944), Paulson (1961), Frye *et al.* (1948, p. 516—Loc. 36), Leonard (1950— Loc. 36), Frye and Leonard (1952—NE¼ Sec. 26, T. 32 s., R. 28 w., Meade Co.).

Locality 4 (Kansas University; Russell County, Kansas)

This locality is the type locality of the *Tobin faunule* (Frye *et al.*, 1943; Hibbard, 1944). The locality is on the east side of the ditch, along the north-south road which is built on the one-half mile line (north-south half section line), and just to the east through the fence in the E½ Sec. 35, T. 14 s., R. 11 w., also just west in the pasture across the road in the W½ of Sec. 35, T. 14 s., R. 11 w., Russell County. The molluscs are listed in Frye *et al.* (1948—Loc. 22), Leonard (1950— Loc. 22); and Frye and Leonard (1952—Loc. NW¼ Sec. 36, T. 14 s., R. 11 w., Russell Co.).

Locality 4 (Kansas University; Lincoln County, Kansas)

This is the type locality of the *Wilson Valley faunule* (Frye *et al.* 1943), located in the roadside ditch and NE corner of SE¼ Sec. 28, T. 13 s., R. 10 w. (reported in 1943 as Sec. 38). The vertebrate fauna was listed by Hibbard (1944), and Guilday and Parmalee (1972) add *Phenacomys*. The molluscs were reported by Frye *et al.* (1948—Loc. 23), Leonard (1950, p. 41—Loc. 23), and Frye and Leonard (1952, p. 159, pl. 8, fig. c—Loc. NW¼ SW¼ Sec. 27, and NE¼ SE¼ Sec. 28, T. 13 s., R. 10 w., Lincoln Co.).

Locality UM-T1-56 (University of Michigan; Baylor County, Texas)
Material of the *Vera faunule* has been obtained from the sw¼ Sec. 152 of Block
A of Buffalo, Bayo, Brazos, and Colorado Railway Company Survey (Getz and
Hibbard, 1965).

Locality UM-T1-57 (University of Michigan; Knox County, Texas)
Material of the *Vera faunule* has been obtained from the southwest corner of the
N¼ Sec. 101 of Block C of the Houston and Texas Central Railroad Company
Survey (Getz and Hibbard, 1965).

Locality UM-T1-58 (University of Michigan; Knox County, Texas)
Material of the *Vera faunule* has been obtained from east of the catchpens, sw¼
SE¼ Sec. 110 of Block C of the Houston and Texas Central Railroad Company
Survey (Getz and Hibbard, 1965).

Systematic Description of the New Ground Squirrel

Order Rodentia
Family Sciuridae

Genus *Spermophilus* F. Cuvier, 1825

Generic Characters
It was impossible to study all forms of living *Spermophilus* and much of the fossil
material is fragmentary. The following are important publications that should be
reviewed in the study of *Spermophilus* and related genera: Black (1963, p. 195);
Bryant (1945); Hall and Kelson (1959, vol. 1, p. 334); Howell (1938, p. 39);
and Stevens (1966, subgenus *Buiscitellus*). The type species is *Mus citellus*
Linnaeus, 1758.

Spermophilus lorisrusselli sp. nov.
Fig. 2

Etymology
Named in honour of Professor Loris S. Russell who has greatly contributed to our
knowledge of the geology and fossils of Canada.

Holotype
University of Michigan, Museum of Paleontology, v61136, adult right mandible
lacking posterior part, but having I_1, P_4–M_3 (Fig. 2B); KU Loc. 4 (Wilson Valley
faunule), late Kansan age, taken just below type O, Pearlette-like volcanic ash
(600,000 m.y.), NE¼ SE¼ Sec. 28, T. 13 s., R. 10 w., Lincoln County, Kansas.
Collected July 31, 1971, by Hibbard and party.

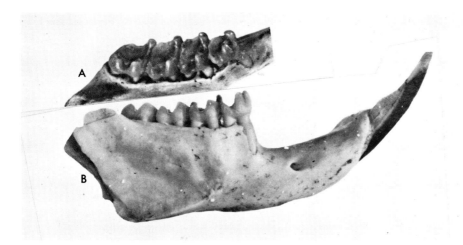

Fig. 2   *Spermophilus lorisrusselli*, sp. nov.
   A. Holotype, right P_4–M_3, UMMP v61136, occlusal view, $\times$ 4.
   B. Holotype, right mandibular ramus with I and P_4–M_3,
      UMMP v61136, buccal view, $\times$ 4.

Specific Characters

The P_4–M_3 are smaller than but similar in pattern to those of *Spermophilus franklinii* (Sabine). An anteroconid (anterior cingulum, paraconulid, and protoconulid of authors) is present on P_4. M_1 has a well-developed pit (trigonid basin) between the protoconid and metaconid (paraconid and parametaconid of some authors). The pit is lacking on M_2. M_2 and M_3 have a narrow metalophid connecting the protoconid and metaconid. P_4 slightly longer than wide; M_1 and M_2 slightly wider than long.

Description

The anteroconid of P_4 joins the protoconid and metaconid and has no anterior valley where it joins the metaconid (Fig. 2A), as also in *Spermophilus franklinii* or *S. richardsonii* Sabine. The ectolophid is poorly developed. A slight swelling occurs anterolabial to the hypoconid and partly closes the valley between the hypoconid and protoconid. Dimensions: anteroposterior, 2.1 mm; transverse, 1.9 mm. Depth of jaw at mid P_4 on lingual side, 5.8 mm.

The right M_1 has a better developed pit between the protoconid and metaconid than observed in Recent *Spermophilus richardsonii*. Dimensions: anteroposterior, 2.25 mm; transverse, 2.35 mm.

The right M_2 has the protoconid connected to the metaconid by a narrow metalophid; no pit is present. Dimensions: anteroposterior, 2.65 mm; transverse, 2.66 mm.

The right M_3 has a narrow metalophid, and a poorly developed ectolophid. It is more triangular in shape than the M_3 of Recent *S. franklinii*. Dimensions: anteroposterior, 3.35 mm; transverse, 2.6 mm.

The M_1–M_3 have no mesoconid or mesostylid. Re-entrant valleys between the protoconids and hypoconids of P_4–M_3 are broader than those of *S. richardsonii*. The length of P_4–M_3 is 9.0 mm.

Remarks

No maxillae or isolated P^3's have been found, so assignment of the fossil ground squirrel to a subgenus is impossible. Seven of the isolated lower molars in v42319 reported by Paulson (1961) as *Citellus* nr. *franklinii* (Sabine) compare well with teeth of the holotype. In the lot are three M_3's that match the shape of M_3 in the holotype. I can see no reason why this species could not have given rise to our modern Franklin Ground Squirrel. Paulson (1961) reported the presence of *Spermophilus richardsonii* and *S.* cf. *S. tridecemlineatus* (Mitchell) from the Cudahy fauna. For a bibliography of this fauna see Hibbard (1970, p. 420).

Acknowledgments

Financial support for the field work in Kansas during the summer of 1971 was provided by the National Science Foundation (project GB-20249). The field crew consisted of Samuel W. Awalt, Robert G. Habetler, Dominique and Thérèse Jammot, Terry L. McMullen, Timothy E. Townsend, and me. Figure 1 was made by Stephen A. Hall. Drs. Emmet T. Hooper and Douglas M. Lay, Mammal Division, The University of Michigan Museum of Zoology, provided Recent osteological specimens for comparison, Mr. Karoly Kutasi photographed the specimen, and Mrs. Gladys Newton typed the manuscript; their help is gratefully acknowledged.

Literature Cited

BLACK, C.C.
   1963   A review of the North American Tertiary Sciuridae. Bull. Mus. Comp. Zool. Harv., 130: 111–248, 8 figs., 22 pls.

BRYANT, M.D.
   1945   Phylogeny of Nearctic Sciuridae. Am. Midl. Nat., 33: 257–390, 48 figs., 8 pls.

CUVIER, F.
   1825   Des dents des Mammifères, considérés comme caractères zoologiques. Paris, iv and 258 pp., 118 pls.

FLINT, R.F.
   1971   Glacial and Quaternary geology. New York, John Wiley and Sons, 892 pp., 230 figs., 75 tables.

FRYE, J.C. AND A.B. LEONARD
1952   Pleistocene geology of Kansas. Kansas Geol. Survey, Bull. 99: 1–230, 17 figs., 19 pls.

FRYE, J.C. AND A.B. LEONARD
1965   Quaternary of the southern Great Plains. *In* Wright, H.E., Jr., and D.G. Frey (eds.), The Quaternary of the United States. Princeton Univ. Press, pp. 203–216, 4 figs.

FRYE, J.C., A.B. LEONARD AND C.W. HIBBARD
1943   Westward extension of the Kansas "*Equus* Beds." Jour. Geol., 51: 33–47, 3 figs., 4 tables.

FRYE, J.C., A. SWINEFORD AND A.B. LEONARD
1948   Correlation of Pleistocene deposits of the central Great Plains with the glacial section. Jour. Geol., 56: 501–525, 3 figs., 2 pls., 2 tables.

GETZ, L.L. AND C.W.HIBBARD
1965   A molluscan faunule from the Seymour Formation of Baylor and Knox counties, Texas. Papers Mich. Acad. Sci., Arts, and Let., 50: 275–297, 1 fig., 1 pl., 3 tables.

GUILDAY, J.E. AND P.W. PARMALEE
1972   Quaternary periglacial records of voles of the genus *Phenacomys* Merriam (Cricetidae: Rodentia). Quaternary Research, 2: 170–175, 2 figs.

HALL, E.R. AND K.R. KELSON
1959   The Mammals of North America. Vol. 1. New York, The Ronald Press, v–xxii and 546 pp., 312 figs., 320 maps.

HIBBARD, C.W.
1944   Stratigraphy and vertebrate paleontology of Pleistocene deposits of southwestern Kansas. Bull. Geol. Soc. Am., 55: 707–754, 20 figs., 3 pls.
1948   Late Cenozoic climatic conditions in the high plains of western Kansas. Bull. Geol. Soc. Am., 59: 592–597, 2 figs.
1949   Pleistocene stratigraphy and paleontology of Meade County, Kansas. Contrib. Mus. Paleo., Univ. Mich., 7: 63–90, 3 figs., 1 pl., 3 maps.
1956   Vertebrate fossils from the Meade Formation of southwestern Kansas. Papers Mich. Acad. Sci., Arts, and Let., 41: 145–203, 16 figs., 2 pls.
1958a   New stratigraphic names for early Pleistocene deposits in southwestern Kansas. Am. Jour. Sci., 256: 54–59, 1 fig.
1958b   Summary of North American Pleistocene mammalian local faunas. Papers Mich. Acad. Sci., Arts, and Let., 43: 3–32, 1 table.
1970   Pleistocene mammalian local faunas from the Great Plains and Central Lowland provinces of the United States. *In* Dort, W., Jr., and J.K. Jones, Jr. (eds.) Pleistocene and Recent environments of the central Great Plains. Dept. Geol. Univ. Press Kansas, Special Publ. 3: 395–433, 1 fig., 8 tables.

HIBBARD, C.W. AND W.W. DALQUEST
1966   Fossils from the Seymour Formation of Knox and Baylor counties, Texas, and their bearing on the late Kansan climate of that Region. Contrib. Mus. Paleo., Univ. Mich., 21: 1–66, 8 figs., 5 pls.

HIBBARD, C.W., C.E. RAY, D.E. SAVAGE, D.W. TAYLOR AND J.E. GUILDAY
1965   Quaternary mammals of North America. *In* Wright, H.E. Jr., and D.G. Frey (eds.), The Quaternary of the United States. Princeton Univ. Press, pp. 509–525, 9 figs., 4 tables.

HOWELL, A.H.
1938   Revision of the North American ground squirrels. North Am. Fauna, 56: 1–256, 20 figs., 32 pls.

IZETT, G.A., R.E. WILCOX, J.D. OBRADOVICH AND R.L. REYNOLDS
    1971   Evidence for two Pearlette-like ash beds in Nebraska and adjoining areas. Geol.
            Soc. Am. abstracts with programs, 3: 265–266.

JAMMOT, D.
    1972   Relationship between the new species of *Sorex scottensis* and the fossil shrews
            Sorex cinereus Kerr. Mammalia, 36: 449–458, 1 fig.

LEONARD, A.B.
    1950   A Yarmouth molluscan fauna in the midcontinent region of the United States.
            Univ. Kansas Paleo. Contrib., Mollusca, art. 3: 1–48, 4 figs., 6 pls.
    1952   Illinoian and Wisconsin molluscan faunas in Kansas. Univ. Kansas Paleo. Contrib.,
            Mollusca, art. 4: 1–38, 15 figs., 5 pls.

LINNAEUS, C.
    1758   Systema naturae [Tenth edition]. Holmiae, vol. 1, 824 pp.

MOORE, R.C., J.C. FRYE, J.M. JEWETT, W. LEE AND H.G. O'CONNOR
    1951   The Kansas rock column. Kansas Geol. Survey. Bull. 89: 1–132, 52 figs.

NAESER, C.W., G.A. IZETT AND R.E. WILCOX
    1971   Zircon fission-track ages of Pearlette-like volcanic ash beds in the Great Plains.
            Geol. Soc. Am., abstracts with programs, 3(7): 657.

PAULSON, G.R.
    1961   The mammals of the Cudahy fauna. Papers Mich. Acad. Sci., Arts, and Let., 46:
            127–153, 8 figs.

SCHULTZ, C.B., L.G. TANNER AND L.D. MARTIN
    1972   Phyletic trends in certain lineages of Quaternary mammals. Bull. Univ. Nebr. St.
            Mus., 9: 183–195, 6 figs.

SKINNER, M.F. AND C.W. HIBBARD, with collaborators
    1972   Early Pleistocene preglacial and glacial rocks and faunas of north-central Nebraska.
            Bull. Am. Mus. Nat. Hist., 148: 1–148, 60 figs., 21 tables.

SMITH, H.T.U.
    1938   Preliminary notes on Pleistocene gravels in southwestern Kansas. Trans. Kansas
            Acad. Sci., 40: 283–291, 1 fig.
    1940   Geologic studies in southwestern Kansas. Kansas Geol. Survey, Bull. 34: 1–212,
            22 figs., 34 pls.

STEVENS, M.S.
    1966   The osteology and relationships of the Pliocene ground squirrel, *Citellus dotti*
            Hibbard, from the Ogallala Formation of Beaver County, Oklahoma. Pearce-
            Sellards Series, 4: 1–24, 6 figs., 3 tables.